Infosys Science Foundation Series

Infosys Science Foundation Series in Mathematical Sciences

The **Infosys Science Foundation Series in Mathematical Sciences**, a Scopus-indexed book series, focuses on high-quality content in the domain of pure and applied mathematics, biomathematics, financial mathematics, operations research and theoretical computer science. With this series, Springer hopes to provide readers with monographs, textbooks, edited volume, handbooks and professional books of the highest academic quality on current topics in relevant disciplines. Literature in this series will appeal to a wide audience of researchers, students, educators and professionals across the world.

Saradha Natarajan

Perfect Powers—An Ode to Erdős

Springer

Saradha Natarajan
Mumbai, Maharashtra, India

ISSN 2363-6149 ISSN 2363-6157 (electronic)
Infosys Science Foundation Series
ISSN 2364-4036 ISSN 2364-4044 (electronic)
Infosys Science Foundation Series in Mathematical Sciences
ISBN 978-981-96-2598-7 ISBN 978-981-96-2599-4 (eBook)
https://doi.org/10.1007/978-981-96-2599-4

Mathematics Subject Classification: 11-xx, 11D61, 11G05, 11B25, 11A41, 11D45, 11D41

This Springer imprint is published by the registered company Springer Nature Singapore Pte Ltd.
The registered company address is: 152 Beach Road, #21-01/04 Gateway East, Singapore 189721, Singapore

Dedicated to T. N. Shorey

Preface

It is well known that a product of two consecutive positive integers is not a square. Goldbach proved that a product of three consecutive positive integers cannot be a square in a letter written to Bernoulli in 1724. It was known to Fermat that there cannot be four squares in arithmetic progression. Euler showed a stronger result that a product of four terms in arithmetic progression cannot be a square. In 1933, Obláth wrote a great deal of the early literature on this folklore problem whether a product of two or more consecutive positive integers is a square or not. More generally, one asks if a product of two or more positive integers in arithmetic progression can be a square or a higher power. This simple question encodes very rich mathematics. My aim in this book is to reveal the various ideas and methods that were developed in an attempt to solve this question. The starting point is a result of Erdős and Rigge in 1939 when they showed independently that a product of two or more consecutive positive integers is never a square. Written as a Diophantine equation this means that the equation

$$x(x+1)\cdots(x+k-1) = y^2 \tag{1}$$

has no solution in integers $x \geq 1, k \geq 2$ and $y \geq 1$. Subsequently, Erdős worked on the equation

$$x(x+1)\cdots(x+k-1) = y^\ell \tag{2}$$

in integers $x \geq 1, k \geq 2, y \geq 1$ and $\ell \geq 2$. In the period between 1951 and 1955, he wrote three papers as a result of which he showed that (2) with $\ell \geq 3$ implies that k is bounded. He applied a method of Thue to show that this equation has only finitely many solutions in x and y. This result is *ineffective* in the sense that x and y cannot be bounded. Using graph theoretical ideas and computations, Erdős and Selfridge completely solved (2) in 1975. Thereafter Erdős made the following *conjecture*. Consider

$$x(x+d)\cdots(x+(k-1)d) = y^\ell \tag{3}$$

in integers $x \geq 1, d \geq 2, k \geq 2, y \geq 1$ and $\ell \geq 2$. Then

Equation (3) *implies that k is bounded by an absolute constant.*

Surprisingly, this conjecture is still *open* in its full generality. Several special cases of the conjecture have been proved since the remarkable paper of Erdős and Selfridge. The late 60s saw the development of the Theory of Linear Forms in logarithms of algebraic numbers by Baker. He demonstrated how his theory can be applied to make results on Thue equations *effective*. Two main contributors of (3) are T. N. Shorey and R. Tijdeman. The former applied Linear Forms in Logarithms for the first time to this equation. Together, Shorey and Tijdeman have written several papers on this equation spanning over two decades. In this book, I will explain and highlight many of their contributions. Toward Erdős conjecture, one of their important contributions is that

(3) *implies that k is bounded by a number depending only on ℓ and $\omega(d)$ where $\omega(d)$ denotes the number of distinct prime divisors of d.*

In particular, their result solves Erdős conjecture when ℓ is fixed and d is prime. They also considered analogous equation to (3) where the product on the left-hand side has many missing terms. Besides the use of Linear Forms in logarithms, hypergeometric methods and various other tools from Analytic Number Theory, their method involves ingenious application of Pigeon hole or Box Principle in a true spirit of Erdős. I will bring out these ideas for a wider audience in this book. A main ingredient is a classical theorem of Sylvester on primes dividing the product on the left-hand side of (3) when $d = 1$ i.e., consecutive integers case. His theorem was extended by Shorey and Tijdeman in the arithmetic progression case, i.e., when $d > 1$. Results from Prime Number theory play an important role here.

In 1993, Wiles solved Fermat's Last Theorem and it has opened the Pandora's box. In the phase between 1995 and 2005 Shorey along with his collaborators N. Saradha (author), S. Laishram and others made use of contributions on Generalized Fermat Equations along with already existing methods to prove new results. As an example, it was shown that there is an infinite set of values of d for which (3) has no solution. I will explain in this book the application of methods from the proof of Fermat's Last Theorem to (3). During this time, Hungarian mathematicians K. Győry, L. Hajdu and Canadian mathematician M. Bennett and others also contributed substantially toward Erdős conjecture which will be traced in this book. So far, all attempts were following classical and semi-classical ideas. After 2005, as a new approach, (3) was looked upon as a superelliptic curve. The problem was reduced to solving ternary Diophantine equations via Frey curves and modular Galois representations. They also depend on results on lower degree equations representing curves of small genus. I shall explain these ideas and apply them to (3).

Despite so many efforts from several mathematicians and application of sophisticated techniques, the conjecture of Erdős is not solved completely. This book will bring together many of the methods that have been used so far and also many of the existing results in one place so that the interested reader will get inspired and try to solve the conjecture one day. When students see various methods in one place, they can understand how the ideas developed. The book is not a compendium of all the results in this area. My aim is not only to give the proofs of some important theorems on (3), but to illustrate the methods and their limitations.

The book has eight chapters. In Chap. 1, some standard results which are basic and necessary for the subsequent chapters are collected. This would enable a smooth flow of proofs in later chapters. Necessary results from Prime Number theory, Baker's theory of Linear Forms in logarithms, Elliptic Curves and Modular forms are presented. In Chap. 2, basic ingredients like the counting arguments and graph theoretic argument of Erdős which go into the proofs are described in detail. These will be used throughout the book. Can a sequence of consecutive integers have an integer which has a large prime divisor? This was answered by Sylvester in the positive. In Chap. 3, proofs of Sylvester's theorem by Erdős and its extension to arithmetic progression by Shorey and Tijdeman are given. Sylvester's theorem is applied to show the irreducibility of truncated exponential polynomials. Some extensions and refinements are also discussed. Chapter 4 deals with a slightly more general result of Erdős and Rigge that a product of two or more consecutive positive integers is never a square. A result of Shorey and Tijdeman regarding squares in a product with terms in an arithmetic progression is proved. Chapter 5 deals with cubes and higher powers in products of consecutive integers and products with terms in an arithmetic progression. A more general result of Erdős and Selfridge for consecutive integers is shown. A result of Shorey and Tijdeman for arithmetic progression together with an improvement by Das-Laishram-Saradha for cubes is proved. Chapter 6 consists of a result of Saradha and Shorey that when a product with terms from an arithmetic progression having common difference $d > 1$ is an ℓ-th power with $\ell \geq 3$, then d has to be really large. As a consequence of a general result, it is shown that $d \neq 2^\alpha 3^\beta 5^\gamma$. In Chap. 7, results on Generalized Fermat Equations are applied to the Erdős' problem. As an illustration, a product of six consecutive positive terms with one term omitted and products with terms in an arithmetic progression of length ≤ 8 are taken and are solved completely. In the last Chap. 8, it is shown that Erdős conjecture which eludes still is a consequence of the ubiquitous *abc*-conjecture. Further we consider the equation from the perspective of a curve and finding rational points on the curve. We prove a result of Bennett and Siksek which is a rational analogue of the famous result of Schinzel and Tijdeman on integral solutions of superelliptic equations. We end the chapter with a very brief discussion on perfect powers in Recurrence sequences, Nagell-Ljunggren equation, Lebesgue-Ramanujan-Nagell equation and Power sums.

I owe this book to my daughter Ambika who inspired me in spite of her difficulties. I thank my daughter Bala to whom I had to give the 'progress report' every now and then. I thank my husband Natarajan and my son-in-law Shivraj and the two l'le uns for showing interest in my work and keeping it lively. On the professional front, I thank Tata Institute of Fundamental Research, Mumbai and in particular Professor Anish Ghosh for giving me access to MATHEMATICA so that I could do all the calculations shown in this book with ease. I thank my young colleagues and collaborators Divyum, Pranabesh and Shanta for all the feedback. My special thanks to Pranabesh for computing the data in Table 7.1 and Divyum for the graphs in

Chap. 2. I thank all my long time collaborators in the area of Diophantine Equations, especially—Profs. Győry, Hajdu and Tijdeman who were very generous to share their ideas with me. Finally my sincere thanks to Shorey who guided me through this beautiful garden of Diophantine Equations sharing and putting ideas in proper perspective.

Mumbai, India Saradha Natarajan

Notations

Let $n \geq 1, d \geq 1$ be integers and x any real number. Let $f(x)$ and $g(x)$ be real valued functions of x with $g(x) \geq 0$. We use the following notations throughout the book.

(1) $2 = p_1 < p_2 < \ldots$—Sequence of all primes.
(2) $P(n)$—Greatest prime factor of n with $P(1) = 1$.
(3) $\omega(n)$—Number of distinct prime factors of n with $\omega(1) = 0$.
(4) $\Omega(n)$—Total number of prime factors of n including their multiplicity with $\Omega(1) = 0$.
(5) $\pi(x)$—Number of primes $\leq x$ for $x > 0$.
(6) $\pi_d(x)$—Number of primes $\leq x$ and coprime to d. When $d = 1$, $\pi_d(x) = \pi(x)$.
(7) $\binom{n}{k} = \frac{n!}{k!(n-k)!}$.
(8) $\nu_p(n) = \text{ord}_p(n) =$ Power of the prime p in the factorization of n.
(9) $x =$ Least integer greater than or equal to x.
(10) $\lfloor x \rfloor =$ Greatest integer less than or equal to x.
(11) We say $f(x) = O(g(x))$ or $f(x) \ll g(x)$ if $|f(x)| \leq cg(x)$ for any real x where c is an absolute constant. When c depends on some parameters, say ϵ, we write $f(x) = O_\epsilon(g(x))$ or $f(x) \ll_\epsilon g(x)$.
(12) We say $|f(x)| = o(g(x))$ if $\lim_{x\to\infty} \frac{|f(x)|}{g(x)} = 0$.
(13) We say $|f(x)| \sim g(x)$ if $\lim_{x\to\infty} \frac{|f(x)|}{g(x)} = 1$.
(14) $\mathbb{N}$—Set of Natural Integers;
$\mathbb{Z}$—Ring of Integers;
$\mathbb{Q}$—Field of Rational Numbers;
$\mathbb{R}$—Field of Real Numbers;
$\mathbb{C}$—Field of Complex Numbers;
(15) If Z is a set, $Z[X]$ denotes the set of polynomials with coefficients in Z.

Contents

Chapter 1
Preliminaries—A Tool Kit

$\prod_{p \leq n} p \leq 4^n$

This chapter is a collection of some results from other areas of Number Theory which are applied in this book. Section 1.1 deals with estimates for prime counting functions $\pi(x)$, $\vartheta(x)$ and factorials. In Sect. 1.2, we state a theorem of Baker (Theorem 1.2.1) giving lower bounds for a linear form in logarithms of rational numbers with rational coefficients and a result of Shorey (Theorem 1.2.2) giving sharper lower bounds. Further we state two results of Baker (Theorems 1.2.3 and 1.2.4) on superelliptic equations and rational approximations to algebraic numbers followed by two applications due to Evertse (Lemma 1.2.5) and Bennett (Theorem 1.2.6). Section 1.3 gives a result from the works of Roth and Halberstam on k-free integers (Lemma 1.3.1). In Sect. 1.4, we record some results of Erdős based on sieve methods. In Sect. 1.5, we show how Erdős used elementary arguments to give upper bounds for products of consecutive primes. The last Sect. 1.6 gives a summary of some basic ideas in the area of Modular forms and important theorems such as Modularity Theorem for Elliptic curves (Theorem 1.6.10) and Ribet's level lowering theorem (Theorem 1.6.12). Further we illustrate their application to Generalized Fermat Equations (Lemmas 1.6.14, 1.6.15 and 1.6.16).

1.1 Results from Prime Number Theory

The Prime Number Theorem which is a fundamental theorem in Number Theory was proposed by Legendre in 1798 and Gauss in 1849. It was proved independently by Hadamard and de la valée Poussin in 1896. It states that

$$\pi(x) \sim \frac{x}{\log x}.$$

S. Natarajan, *Perfect Powers—An Ode to Erdős*, Infosys Science Foundation Series,
https://doi.org/10.1007/978-981-96-2599-4_1

Thus

$$\lim_{x\to\infty} \frac{\pi(x)}{x/\log x} = 1 \text{ i.e., } \pi(x) - \frac{x}{\log x} = o\left(\frac{x}{\log x}\right).$$

The problem of improving the error term $o\left(\frac{x}{\log x}\right)$ is considered by many mathematicians, see the paper of Rosser and Schoenfeld [1]. For many applications, this is desirable. We quote below the result due to Dusart [2]. These are two good sources for many inequalities on arithmetical functions involving primes.

Theorem 1.1.1 *For $x > 1$, we have*

$$\pi(x) \leq \frac{x}{\log x} + \frac{1.2762x}{\log^2 x}.$$

On the other side, we also have by [1, 2] that

$$\pi(x) \begin{cases} > x/\log x \text{ for } x \geq 17 \\ \geq x/(\log x - 1) \text{ for } x \geq 5393. \end{cases} \tag{1.1}$$

We shall list a few more resuts from Prime Number theory.

$$\vartheta(x) < 1.01624x \tag{1.2}$$

where

$$\vartheta(x) = \sum_{\substack{p\leq x \\ p-\text{prime}}} \log p.$$

$$p_n > n \log n \tag{1.3}$$

where p_n is the nth prime. Next consider $\omega(n)$, the number of distinct prime divisors of $n > 1$. It is easy to see that $\omega(n) \leq \frac{\log n}{\log 2}$. With a little more effort it can be seen that

$$\omega(n) \ll \frac{\log n}{\log \log n}. \tag{1.4}$$

One can show that the constant involved in $\ll$ above is ≤ 4.

Next we state the standard Stirling's formula for factorials. For any integer $M > 1$, we have

$$\log M! < \log \sqrt{2\pi} + \left(M + \frac{1}{2}\right) \log M - M + \frac{1}{12M}, \tag{1.5}$$

$$\log M! > \log \sqrt{2\pi} + \left(M + \frac{1}{2}\right) \log M - M. \tag{1.6}$$

1.2 Results from the Theory of Linear Forms in Logarithms

In the late sixties and early seventies, Baker established his profound theory on Linear Forms in Logarithms of algebraic numbers. We refer to the books of [3–5] for details on this theory and its applications to various areas of Number Theory. We shall present two results which will be applied in this book. The first one is an estimate of Baker [6] on Linear Forms in Logarithms of rational numbers. The height of a rational number u/v with $\gcd(u, v) = 1$ is defined as $\max(|u|, |v|)$.

Theorem 1.2.1 *Let $\alpha_1, \ldots, \alpha_n$ be positive rational numbers of heights not exceeding $A_1, \ldots, A_n$, respectively, where $A_j \geq 3$ for $1 \leq j \leq n$. Put*

$$\Omega' = \prod_{j=1}^{n-1} \log A_j, \quad \Omega = \Omega' \log A_n.$$

There exist effectively computable positive absolute constants C_1 and C_2 such that

$$\left|b_1 \log \alpha_1 + \cdots + b_n \log \alpha_n\right| > \exp\left(-(C_1 n)^{C_2 n} \Omega \log \Omega' \log B\right)$$

for all rational integers $b_1, \ldots, b_n$ of absolute values not exceeding $B(\geq 3)$.

In 1986, Shorey [7, Lemma 2] showed that sharper estimates for Linear Forms in Logarithms can be obtained if α_i's are very close to 1. The result is as follows.

Theorem 1.2.2 *Let $\alpha_1, \ldots, \alpha_{n-1}$ be positive rational numbers of heights $\leq A_1(\geq 3)$, and α_n a rational number of height $\leq A(\geq 3)$ with*

$$\frac{\log A}{\log A_1} \geq \frac{1}{\tau_1} \text{ where } \tau_1 \geq 1$$

and

$$\left|\log \alpha_i\right| \leq A_1^{-1/\tau_2} \text{ for } 1 \leq i \leq n \text{ with } \tau_2 \geq 1.$$

Let $b_1, \ldots, b_n$ be rational integers of absolute value $\leq A_1^u$. Then there exist effectively computable positive numbers C_3 and C_4 depending only on u and a prime $p \leq C_3 \tau_2 n^3$ such that the inequality

$$\left|b_1 \log \alpha_1 + \cdots + b_n \log \alpha_n\right| < \exp\left(-(C_4 \tau_2 n^3)^{3n+3} \tau_1 \log A\right)$$

implies that

$$\alpha_1^{j_1} \cdots \alpha_n^{j_n} = \eta^p$$

where η is a rational number and $j_1, \ldots, j_n$ are integers satisfying $0 \leq j_i < p$ for $1 \leq i < n$ and $j_n = 1$.

As one of the consequences of his theory, Baker (see [8]) proved the following theorem on superelliptic equations. See also [4, Chap. 6]

Theorem 1.2.3 *Let $f(X) \in \mathbb{Z}[X]$. Assume that the equation*

$$f(x) = y^m$$

holds in rational integers x, y, m with $m \geq 2$. Further assume that $f(X)$ has at least two simple roots if $m \geq 3$ and three simple roots if $m = 2$. Then

$$\max\left(|x|, |y|\right) \leq C(m, f)$$

where $C(m, f)$ is an effectively computable number depending only on m and f.

We record here another result of Baker [9] on rational approximations to algebraic numbers wherein he established completely explicit results for certain class of algebraic numbers.

Theorem 1.2.4 *Suppose that m, ℓ are integers such that*

$$\ell \geq 3 \text{ and } 1 \leq m < \ell.$$

Let a, b be positive integers for which $\frac{7}{8}a \leq b < a$ and suppose that $a - b$ is divisible by ℓ. Suppose also that

$$\mu_\ell = \prod_{\substack{p|\ell \\ p-prime}} p^{1/(p-1)}, \ \lambda = 4b(a-b)^{-2}\mu_\ell^{-1} > 1.$$

Then the following inequality

$$\left| \left(\frac{a}{b}\right)^{m/\ell} - \frac{p}{q} \right| > \frac{c}{q^\kappa} \tag{1.7}$$

holds for all integers p and $q > 0$ where κ and c are given by

$$\lambda^{\kappa-1} = 2\mu_\ell(a+b); \ c^{-1} = 2^{\kappa+2}(a+b).$$

As a consequence of Baker's theory of Linear Forms in Logarithms, many Diophantine equations could be solved effectively in the sense that the variables involved could be bounded. One such class of equations goes by the name *Thue equations.* In 1982, Evertse [10] dealt with *diagonal* Thue equations viz., $aX^\ell - bY^\ell = c$. We need the following result from his paper [10, p. 290]. Let $R(\ell, d)$ denote the number of residue classes z such that $z^\ell \equiv 1 \pmod{d}$. Note that $R(\ell, d)$ is a multiplicative function of d. Let

$$d = 2^\alpha d' \text{ where } \alpha \text{ is a non negative integer}$$

and

$$d' = q_1^{\alpha_1} \cdots q_t^{\alpha_t}$$

with $q_1, \cdots, q_t$ distinct odd primes and $\alpha_1, \ldots, \alpha_t$ positive integers. Then

$$R(\ell, d) = R(\ell, 2^{\alpha})R(\ell, d') = R(\ell, 2^{\alpha}) \prod_{i=1}^{t} R(\ell, p_i^{\alpha_i}).$$

Further

$$R(\ell, p_i^{\alpha_i}) = \gcd\left(\ell, \varphi(p_i^{\alpha_i})\right), 1 \le i \le t.$$

So $R(\ell, d')$ divides

$$\gcd\left(\ell^t, \prod_{i=1}^{t} \varphi(p_i^{\alpha_i})\right) = \gcd\left(\ell^{\omega(d')}, \varphi(d')\right).$$

Also $R(\ell, 2^{\alpha}) = 1$ if $\alpha \in \{0, 1\}$ and $R(\ell, 2^{\alpha}) = \gcd(\ell, 2)\gcd(\ell, 2^{\alpha-2})$ if $\alpha \ge 2$. (see [11, Corollary 2.44]). Hence $R(\ell, 2^{\alpha})$ divides $\gcd\left(\gcd(\ell, 2)\ell, \varphi(2^{\alpha})\right)$. Thus $R(\ell, d)$ divides $\gcd\left(\gcd(\ell, 2)\ell^{\omega(d)}, \varphi(d)\right)$. So

$$R(\ell, d) \le \gcd(\ell, 2)\ell^{\omega(d)}.$$

The above inequality implies the following lemma.

Lemma 1.2.5 *Let ℓ be an odd prime number. Let $R(\ell, d)$ denote the number of residue classes z such that $z^{\ell} \equiv 1 \pmod{d}$. Then $R(\ell, d) \le \ell^{\omega(d)}$.*

Another important result on Thue equations was due to Bennett [12, Theorem 1.1] which goes as follows.

Theorem 1.2.6 *If a, b and ℓ are integers with $ab \ne 0$ and $\ell \ge 3$, then the equation*

$$|ax^{\ell} - by^{\ell}| = 1$$

has at most one solution in positive integers (x, y). Hence the equation

$$|(a+1)x^{\ell} - ay^{\ell}| = 1$$

has the only solution $(x, y) = (1, 1)$.

1.3 Results on k-free Integers

In the Number System, following the primes on the heels are square free integers and k-free integers in general. One of the well analyzed questions is about the gap between squarefree numbers or k-free numbers. For instance, it is well known that the number of squarefree numbers $\leq X$ is

$$\frac{6}{\pi^2}X + O(X^{1/2}),$$

so that

$$q_2(n+1) - q_2(n) = O(n^{1/2})$$

where $q_k(n)$ denotes the nth k-free number. In 1951, Roth [13], Halberstam and Roth [14] made important contributions to this topic. By a simple elementary argument (alluded to Davenport), it is shown in [13] that

$$q_2(n+1) - q_2(n) = O(n^{1/3}(\log n)^{-2/3}).$$

Using analytic methods Roth improved this as

$$q_2(n+1) - q_2(n) = O(n^{3/13}(\log n)^{-4/13}).$$

An extension of a simple and elementary argument for square free integers due to Estermann gives

$$q_k(n+1) - q_k(n) = O(n^{1/(k+1)}).$$

Halberstam and Roth showed that

$$q_k(n+1) - q_k(n) = O(n^{1/(2k)+\epsilon}).$$

We are interested in [14, Lemma 2] which is as below.

Lemma 1.3.1 *Suppose that*

$$A\left|X - X_0\right|^{2\ell-1} \leq \frac{1}{2}X_0^{\ell}, \tag{1.8}$$

and

$$\left|AX^{\ell} - A_0X_0^{\ell}\right| \leq \eta X_0, \tag{1.9}$$

where η is sufficiently small as a function of ℓ. Then for any given (A_0, X_0), the number of pairs (A, X) satisfying (1.8) and (1.9) is $O\big((A_0X_0)^{\epsilon}\big)$ for any $\epsilon > 0$.

1.4 Sieve Theoretic Arguments of Erdős

Erdős [15] introduced some applications of Brun's sieve for dealing with problems considered in this book. We give here some results used in the book. For understanding sieve methods, see [16]. The following lemma is taken from [16, p. 68].

Lemma 1.4.1 *Let $X > e$. Let $\Phi(X, w)$ denote the number of integers $\leq X$ free of prime factors $\leq w$ (i.e., integers whose prime factors are $> w$). Suppose $w \leq X^{\theta}, 0 < \theta < 1$. Then*

$$\Phi(X, w) \ll_{\theta} X \prod_{p \leq w} \left(1 - \frac{1}{p}\right).$$

It is well known that

$$\prod_{p \leq w} \left(1 - \frac{1}{p}\right) \ll \frac{1}{\log w}.$$

Hence

$$\Phi(X, w) \ll_{\theta} \frac{X}{\log w} \text{ for } w \leq X^{\theta}. \tag{1.10}$$

Erdős applied the above lemma to get the following result.

Lemma 1.4.2 *Let $B = \{b_1, \ldots, b_s\}$ denote the set of all integers between z and X such that every proper divisor of b_i is at most z. Then*

$$s = |B| \ll \frac{X}{\log(X/z)}.$$

Proof Note that all primes between z and X belong to B. Suppose $b \in B$ is composite. Let p be a prime dividing b and write $b = pv$. Then $p \leq z$ and $v \leq z$. The number of choices for p is $\pi(z)$. So the number of such b's is

$$\ll \pi(z)z$$

and the total number of b's is

$$\ll \pi(X) + \pi(z)z.$$

Suppose $z \leq \sqrt{X}$. Then the total number of b's is

$$\ll \frac{X}{\log X}.$$

Since $\frac{X}{\log X/z} \geq \frac{X}{\log X}$, the lemma is true. So let us assume that $z > \sqrt{X}$. Put

$$y = \frac{X}{\log(X/z)}.$$

Take a $b_i > y$. Suppose $b_i = pc_i$ with p prime. Then $p > y/c_i > y/z$ by the property of b_i's. Thus every prime divisor of such b_i exceeds y/z. Note that

$$y/z \leq X^{1/2}.$$

Applying (1.10), the number of such b_i' s is $\ll \frac{X}{\log(y/z)}$. Number of b_i' s $\leq y$ is at most y. Hence

$$|B| \leq \frac{X}{\log(X/z)} + \frac{X}{\log(y/z)}.$$

But $\log(y/z) = \log(X/z) - \log\log(X/z) \geq \frac{1}{2}\log(X/z)$ which together with the above inequality gives the result. □

We apply the above result to find in a sequence, some numbers with large common divisor.

Lemma 1.4.3 *Let $0 < z < X$ with $X > e$. Let $0 < a_1 < a_2 < \cdots < a_r \leq X$ be a sequence of integers. Then there exists an absolute constant c_1 such that there are at least*

$$r - z - \frac{c_1 X}{\log(X/z)}$$

pairs (a_i, a_j) for which $\gcd(a_i, a_j) > z$.

Proof Let the set B be as in Lemma 1.4.2 so that

$$|B| \leq \frac{c_1 X}{\log(X/z)}.$$

Take an integer n with $z < n \leq X$. Suppose $n \notin B$. Then it has a divisor $n_1 > z$. If $n_1 \notin B$ then again there is a divisor $n_2 > z$ of n_1. Proceeding like this we end up with a divisor of n in B. Thus every integer between z and X has a divisor in B. In the given sequence of integers $a_1, \ldots, a_r$ there are at least $r - z$ integers which are $> z$. Let $t_i \geq 0$ be the number of integers from the sequence which are divisible by b_i. Note that $t_i = 1$ for at most $|B|$ of the a_i' s exceeding z. For $t_i \geq 2$, there are $\binom{t_i}{2}$ pairs (a_i, a_j) which have gcd $\geq b_i$. In all, there are

$$\sum_{t_i \geq 2} \binom{t_i}{2} \geq \sum_{t_i \geq 2} t_i \geq r - z - |B|$$

pairs (a_i, a_j) which has some $b \in B$ as gcd which means that there are at least $r - z - |B|$ pairs (a_i, a_j) having gcd $> z$. □

Lemma 1.4.4 *Let $X > e$. Let $0 < a_1 < a_2 < \cdots < a_s \le X$ be all the integers in the interval $(0, X]$ such that the products $a_i a_j$, $1 \le i, j \le s$ are all distinct. Then*

$$s < 2X/\log X$$

for X sufficiently large.

Proof Let the hypotheses of the lemma hold. Write

$$a_i = u_i v_i,\ 1 \le i \le s$$

where u_i is the largest divisor of a_i which is not greater than $X^{1/2}$. Note that for any two a_i and a_j, at least one of

$$u_i v_j \text{ and } u_j v_i$$

is not one of the a' s, since otherwise, we would have $u_i v_j = a'$ and $u_j v_i = a''$ which gives

$$a_i a_j = a' a''$$

a contradiction. (Later we will interpret this property as a certain bipartite graph not having any rectangle). Let us fix a number $v \le X$. Suppose an $a_i = u_i v$ we re-write it as $a_i^{(v)} = u_i^{(v)} v$. Let the set

$$A^{(v)} = \{a_1^{(v)}, \ldots, a_{r_v}^{(v)}\}$$

denote all such a_i's and

$$U^{(v)} = \{u_i^{(v)},\ 1 \le i \le r_v\}.$$

Note that $s = \sum_v r_v$. By the given hypothesis, we find that

$$|U^{(v_1)} \cap U^{(v_2)}| \le 1 \text{ for } v_1 \ne v_2.$$

and all the pairs

$$(u_i^{(v)}, u_j^{(v)}),\ 1 \le i, j \le r_v$$

for any given v are distinct.

Case 1. Suppose in $A^{(v)}$, for some i, $u_i^{(v)} < X^{1/4}$. Let us call such an $A^{(v)}$ as $B^{(v)}$. We show that v is a prime. Suppose not. Let p be the least prime factor of v. Then $p < X^{1/2}$. Since $u_i^{(v)} p$ is a divisor of $a_i^{(v)}$, by the maximality property of $u_i^{(v)}$ we get

$$p > X^{1/2}/u_i^{(v)} > X^{1/4} > u_i^{(v)}.$$

This means $X^{1/2} > v > u_i^{(v)}$ which again contradicts the maximality of $u_i^{(v)}$. This shows that v is a prime. We consider two possibilities.

(i) Suppose $|B^{(v)}| = 1$ for every $v \le X$. Then

$$\sum_v |B^{(v)}| \le \pi(X).$$

(ii) Suppose $|B^{(v)}| > 1$ for some $v \le X$. All the pairs $(u_i^{(v)}, u_j^{(v)})$, $1 \le i < j \le r_v$ are distinct and there are $\binom{r_v}{2}$ of them. As v varies, the total number of such pairs is $\sum_v \binom{r_v}{2}$. On the other hand, since each $u_i^{(v)} < X^{1/4}$, the number of pairs $(u_i^{(v)}, u_j^{(v)})$ is $< X^{1/2}$. Thus

$$\sum_v \binom{r_v}{2} < X^{1/2} \text{ or } \sum_v r_v < 2X^{1/2}.$$

So

$$\sum_v |B^{(v)}| \le 2X^{1/2}.$$

Case 2. Suppose $X^{1/4} \le u_i^{(v)} \le X^{1/2}$ for every $u_i^{(v)} \in A^{(v)}$. Let us call such sets as $C^{(v)}$. Then $v \le X^{3/4}$. There are two possibilities.

(iii) Suppose $r_v \le X^{1/8}$. Then

$$\sum_v |C^{(v)}| \le (X^{1/8} + 1)X^{3/4} \le 2X^{7/8}.$$

(iv) Suppose $r_v > X^{1/8}$. The total number of pairs $(u_i^{(v)}, u_j^{v)})$, $1 \le i, j \le r_v$, $v \le X$ is at most

$$\binom{\lfloor X^{1/2} \rfloor}{2} < X/2,$$

so that $\sum_v \binom{r_v}{2} < X/2$. Since $r_v > X^{1/8}$, we get

$$\sum_v r_v < X^{7/8}.$$

Thus we have

$$\sum_v |C^{(v)}| < X^{7/8}.$$

Summing (i)–(iv), we get

$$N(X) < \pi(X) + 3X^{7/8} + 2X^{1/2}.$$

This proves the lemma. □

1.5 On Primes Dividing Binomial Coefficients

Lemma 1.5.1 *Let p be a prime. Suppose p^α divides $\binom{n}{k}$, then $p^\alpha \le n$.*

Proof Observe that

$$\alpha \le \nu_p\left(\binom{n}{k}\right).$$

So it is enough to show that

$$\nu_p\left(\binom{n}{k}\right) \le \frac{\log n}{\log p}.$$

It is well known that if p divides $m!$ for any positive integer m, then

$$\nu_p(m!) = \left\lfloor \frac{m}{p} \right\rfloor + \left\lfloor \frac{m}{p^2} \right\rfloor + \cdots + \left\lfloor \frac{m}{p^h} \right\rfloor$$

where h is the highest power of p such that $p^h \le m$, i.e., $p^h \le m < p^{h+1}$. Using the above fact,

$$\nu_p\left(\binom{n}{k}\right) = \left(\left\lfloor \frac{n}{p} \right\rfloor - \left\lfloor \frac{k}{p} \right\rfloor - \left\lfloor \frac{n-k}{p} \right\rfloor\right) + \left(\left\lfloor \frac{n}{p^2} \right\rfloor - \left\lfloor \frac{k}{p^2} \right\rfloor - \left\lfloor \frac{n-k}{p^2} \right\rfloor\right)$$
$$+ \cdots + \left(\left\lfloor \frac{n}{p^\beta} \right\rfloor - \left\lfloor \frac{k}{p^\beta} \right\rfloor - \left\lfloor \frac{n-k}{p^\beta} \right\rfloor\right) \tag{1.11}$$

where

$$p^\beta \le n < p^{\beta+1}.$$

Observe that each of these β terms has the value 0 or 1. Thus

$$\nu_p\left(\binom{n}{k}\right) \le \beta \le \frac{\log n}{\log p}$$

which proves the lemma. □

Lemma 1.5.2 *Let $n \geq 1$ be an integer. Let*

$$a_i = \left\lceil \frac{n}{2^i} \right\rceil, i \geq 1.$$

Let m be the smallest integer such that $n/2^m \leq 1$. Then

$$\binom{2a_1}{a_1}\binom{2a_2}{a_2}\cdots\binom{2a_m}{a_m} < 4^n.$$

Proof One can easily check the assertion for small integers, say $n \leq 10$. For instance, when $n = 10$, $a_1 = 5$, $a_2 = 3$, $a_3 = 2$ and $a_4 = 1$. Then

$$\binom{10}{5}\binom{6}{3}\binom{4}{2}\binom{2}{1} = 60480 < 1048576 = 4^{10}.$$

From now on we assume that $n > 10$. Also we check by induction that

$$\binom{2n}{n} < 4^{n-1}. \tag{1.12}$$

Suppose the assertion of the lemma is true for any integer $< n$. We show that the result is true for n. First note that $a_m = 1$ and

$$n > a_1 > a_2 > \cdots > a_m = 1.$$

Also

$$a_1 = \begin{cases} n/2 \; if \; n \text{ is even} \\ (n+1)/2 \; if \; n \text{ is odd} \end{cases}$$

and

$$2a_2 - 1 = \begin{cases} (n-2)/2 \text{ if } n \equiv 0 \pmod 4 \\ (n+1)/2 \text{ if } n \equiv 1 \pmod 4 \\ n/2 \text{ if } n \equiv 2 \pmod 4 \\ (n-1)/2 \text{ if } n \equiv 3 \pmod 4. \end{cases}$$

So we get

$$2a_2 - 1 \leq a_1 \leq (n+1)/2 < n.$$

Let

$$b_i = \left\lceil \frac{2a_2 - 1}{2^i} \right\rceil \text{ for } i \geq 1.$$

Note that $b_1 = a_2$ and for $i \geq 2$,

$$b_i = \left\lceil \frac{a_2}{2^{i-1}} \right\rceil = \left\lceil \frac{\lceil n/2^2 \rceil}{2^{i-1}} \right\rceil \geq \left\lceil \frac{n}{2^{i+1}} \right\rceil = a_{i+1}.$$

Thus

$$\binom{2b_1}{b_1} = \binom{2a_2}{a_2} \text{ and } \binom{2b_i}{b_i} \geq \binom{2a_{i+1}}{a_{i+1}} \text{ for } i \geq 2.$$

By the assertion of the lemma with n replaced by $2a_2 - 1$, we get

$$\binom{2b_1}{b_1}\binom{2b_2}{b_2}\cdots\binom{2b_m}{b_m} < 4^{2a_2-1}.$$

This then gives

$$\binom{2a_2}{a_2}\cdots\binom{2a_m}{a_m} < 4^{2a_2-1}.$$

Now we apply the above inequality with (1.12) and $2a_2 - 1 \leq a_1$ to get

$$\begin{aligned}\binom{2a_1}{a_1}\binom{2a_2}{a_2}\cdots\binom{2a_m}{a_m} &< \binom{2a_1}{a_1}4^{2a_2-1} \\ &< 4^{a_1-1}4^{2a_2-1} = 4^{a_1+2a_2-2} \\ &\leq 4^{2a_1-1} \leq 4^n\end{aligned}$$

which proves the lemma by induction. □

Remark Note that

$$(1, n] = \cup_{i=1}^{m}(a_i,\ 2a_i]$$

and

$$\left(1, \lfloor n^{1/\alpha} \rfloor\right] = \cup_{i=1}^{m}\left(\left\lfloor a_i^{1/\alpha} \right\rfloor, \left\lfloor (2a_i)^{1/\alpha} \right\rfloor\right],$$

i.e., the interval $(1, n]$ is covered by $(a_1, 2a_1], (a_2, 2a_2], \ldots, (a_m, 2a_m]$ and the interval $\left(1, \lfloor n^{1/\alpha} \rfloor\right]$ is covered by

$$\left(\left\lfloor a_1^{1/\alpha} \right\rfloor, \left\lfloor (2a_1)^{1/\alpha} \right\rfloor\right], \ldots, \left(\left\lfloor a_m^{1/\alpha} \right\rfloor, \left\lfloor (2a_m)^{1/\alpha} \right\rfloor\right].$$

Now we analyze the prime divisors of $\binom{2m}{m}$. Surely, any prime p with $m < p \leq 2m$ divides $\binom{2m}{m}$ since the numerator is divisible by such a prime p and the denominator is not. What about prime powers? Let us consider primes satisfying

$$m^{1/\alpha} < p \leq (2m)^{1/\alpha} \text{ for any integer } \alpha \geq 2. \tag{1.13}$$

By (1.11),

$$\nu_p\left(\binom{2m}{m}\right) = \left(\left\lfloor\frac{2m}{p}\right\rfloor - 2\left\lfloor\frac{m}{p}\right\rfloor\right) + \cdots + \left(\left\lfloor\frac{2m}{p^{\alpha-1}}\right\rfloor - 2\left\lfloor\frac{m}{p^{\alpha-1}}\right\rfloor\right) + 1 \geq 1$$

since $\left\lfloor\frac{2m}{p^\alpha}\right\rfloor = 1$ and $\left\lfloor\frac{m}{p^\alpha}\right\rfloor = 0$. Thus every prime in (1.13) also divides $\binom{2m}{m}$. So

$$\prod_{m<p\leq 2m} p \prod_{\lfloor m^{1/2}\rfloor < p \leq \lfloor (2m)^{1/2}\rfloor} p \cdots \leq \binom{2m}{m}.$$

In fact the left-hand side expression divides $\binom{2m}{m}$. Taking $m = a_1, a_2, \ldots, a_m$ and applying the Remark above and Lemma 1.5.2, we conclude that

Theorem 1.5.3 *Let $n > 1$. Then*

$$\prod_{p\leq n} p \prod_{p\leq n^{1/2}} p \prod_{p\leq n^{1/3}} p \cdots < 4^n.$$

In particular,

$$\prod_{p\leq n} p < 4^n.$$

Remark The above inequality is achieved using elementary arguments of Erdős [17]. If one chooses to use (1.2), we get

$$\prod_{p\leq n} p \leq (2.78)^n. \tag{1.14}$$

As a corollary, we derive

Corollary 1.5.4 *Let $k^{3/2} > m \geq k$ be integers. Suppose $\binom{m}{k}$ is not divisible by any prime $> k$. Then*

$$\binom{m}{k} < 4^{k+\sqrt{m}}.$$

Proof Apply Theorem 1.5.3 with $n = k$ and $n = \sqrt{m}$ to get

$$\prod_{p\leq k} p \prod_{p\leq k^{1/2}} p \prod_{p\leq k^{1/3}} p \cdots < 4^k \tag{1.15}$$

and

$$\prod_{p\leq \sqrt{m}} p \prod_{p\leq \sqrt[4]{m}} p \prod_{p\leq \sqrt[6]{m}} p \cdots < 4^{\sqrt{m}}. \tag{1.16}$$

Since $k > m^{2/3}$, we have

$$k^{1/\ell} \geq m^{1/(2\ell-1)} \text{ for } \ell \geq 2.$$

So from (1.15) we get

$$\prod_{p\le k} p \prod_{p\le k^{1/3}} p \prod_{p\le k^{1/5}} p \cdots < 4^k$$

which together with (1.16) implies that

$$\prod_{p\le k} p \prod_{p\le m^{1/2}} p \prod_{p\le m^{1/3}} p \cdots < 4^{k+\sqrt{m}}.$$

Now observe that by Lemma 1.5.1, if p^α divides $\binom{m}{k}$, then $p^\alpha \le m$. So $p \le m^{1/\alpha}$. For $\alpha \ge 2$, $p \le m^{1/\alpha} \le m^{1/2}$. Thus

$$\begin{aligned}\binom{m}{k} &\le \prod_{\substack{p\le k\\ \alpha=\nu_p\binom{m}{k}}} p^\alpha\\ &\le \prod_{p\le k} p \prod_{p\le k^{1/2}} p \prod_{p\le k^{1/3}} p \cdots\\ &\le \prod_{p\le k} p \prod_{p\le m^{1/2}} p \prod_{p\le m^{1/3}} p \cdots\\ &< 4^{k+\sqrt{m}}.\end{aligned}$$

□

1.6 Modular Forms and Elliptic Curves

These wonderful topics saw leaping progress since twentieth century. Wile's proof in 1994 of Fermat's Last Theorem rests heavily on the theory of modular forms and Galois Representations. There are plenty of books available on these topics. The reader may look at Knapp [18], Lang [19], Ogg [20, 21], Silverman [22, 23], Siksek [24] and Shimura [25] to name a few. We also recommend a recent book by Dujella [26]. There are also tables of Cremona [27] for elliptic curves and Stein's [28] modular forms database. In this section, we provide a very brief account of some notations, definitions and some important theorems. We shall apply them to solve some ternary equations which includes Fermat's Last Theorem. This is just to give the reader a quick glance at the machinary involved. The interested reader should consult books on this topic and original papers.

Let R denote any commutative ring with identity. $SL_2(R)$ denotes the group of matrices

$$\begin{pmatrix} a & b\\ c & d\end{pmatrix}$$

of determinant 1 and $a, b, c, d \in R$. We now specialize $R = \mathbb{Z}$. Let

$$\Gamma(N) = \left\{ \begin{pmatrix} a & b \\ c & d \end{pmatrix} \in SL_2(\mathbb{Z}) \middle| \begin{pmatrix} a & b \\ c & d \end{pmatrix} \equiv \begin{pmatrix} 1 & 0 \\ 0 & 1 \end{pmatrix} \pmod{N} \right\}.$$

$\Gamma(N)$ is a normal subgroup of finite index in $SL_2(\mathbb{Z})$ and is called the *principal congruence subgroup* of level N. The *Hecke subgroup* $\Gamma_0(N)$ of level N is given by

$$\Gamma_0(N) = \left\{ \begin{pmatrix} a & b \\ c & d \end{pmatrix} \in SL_2(\mathbb{Z}) \middle| N \text{ divides } c \right\}.$$

Note that

$$\Gamma(N) \subseteq \Gamma_0(N) \subseteq SL_2(\mathbb{Z}).$$

Thus $\Gamma_0(N)$ has finite index in $SL_2(\mathbb{Z})$. Let

$$\Gamma_1(N) = \left\{ \begin{pmatrix} a & b \\ c & d \end{pmatrix} \in SL_2(\mathbb{Z}) \middle| \begin{pmatrix} a & b \\ c & d \end{pmatrix} \equiv \begin{pmatrix} 1 & * \\ 0 & 1 \end{pmatrix} \pmod{N} \right\}.$$

Then

$$\Gamma(N) \subseteq \Gamma_1(N) \subseteq \Gamma_0(N) \subseteq SL_2(\mathbb{Z}).$$

A *congruence subgroup* of $SL_2(\mathbb{Z})$ is defined as a subgroup which contains $\Gamma(N)$ for some N. $\Gamma_0(N)$ and $\Gamma_1(N)$ are examples of congruence subgroups.

Let $\mathcal{H}$ denote the upper half plane. Let $GL_2(\mathbb{R})$ denote the group of 2×2 matrices with real entries and $GL_2^+(\mathbb{R})$, its subgroup with positive determinant. Then $GL_2^+(\mathbb{R})$ acts on $\mathcal{H}$ as a group of automorphisms as follows:

$$\gamma : z \to \frac{az+b}{cz+d}, \ \gamma = \begin{pmatrix} a & b \\ c & d \end{pmatrix} \in GL_2^+(\mathbb{R}).$$

In particular, $SL_2(\mathbb{Z})$ acts on $\mathcal{H}$. Let

$$\mathcal{H}^* = \mathcal{H} \cup \mathbb{Q} \cup \infty.$$

The action of $SL_2(\mathbb{Z})$ can be extended to $\mathcal{H}^*$ by defining

$$\begin{pmatrix} a & b \\ c & d \end{pmatrix} (\infty) = \frac{a}{c} \text{ if } c \neq 0$$

and

$$\begin{pmatrix} a & b \\ 0 & d \end{pmatrix} (\infty) = \infty;$$

for $r/s \in \mathbb{Q}$ with $\gcd(r, s) = 1$, we define

$$\begin{pmatrix} a & b \\ c & d \end{pmatrix} \left(\frac{r}{s}\right) = \begin{cases} \frac{ar+bs}{cr+ds} \text{ if } cr+ds \neq 0 \\ \infty \text{ if } cr+ds = 0. \end{cases}$$

The rational numbers together with ∞ are called *cusps*.

If Γ is a discrete subgroup of $SL_2(\mathbb{R})$, then the orbit space $\mathcal{H}^*/\Gamma$ can be given the structure of a compact Riemann surface and is denoted by X_Γ. We are interested in Γ being a congruence subgroup of $SL_2(\mathbb{Z})$. In this case, the algebraic curve corresponding to X_Γ is called a *modular curve*. When $\Gamma = \Gamma(N), \Gamma_1(N)$ or $\Gamma_0(N)$, the corresponding modular curve is denoted by $X(N)$, $X_1(N)$ and $X_0(N)$, respectively.

Let f be a holomorphic function on $\mathcal{H}$ and k a positive integer. For

$$\gamma = \begin{pmatrix} a & b \\ c & d \end{pmatrix} \in GL_2^+(\mathbb{R}),$$

define

$$(f|_k\gamma)(z) = (det\gamma)^{k/2}(cz+d)^{-k} f\left(\frac{az+b}{cz+d}\right).$$

For fixed k, the map $\gamma : f \to f|_k\gamma$ defines an action of $GL_2^+(\mathbb{R})$ on the space of holomorphic functions on $\mathcal{H}$. Let Γ be a subgroup of finite index in $SL_2(\mathbb{Z})$. Let f be a holomorphic function on $\mathcal{H}$ such that $f|_k\gamma = f$ for all $\gamma \in \Gamma$. Since Γ has finite index,

$$\begin{pmatrix} 1 & 1 \\ 0 & 1 \end{pmatrix}^M = \begin{pmatrix} 1 & M \\ 0 & 1 \end{pmatrix} \in \Gamma$$

for some positive integer M. Hence $f(z+M) = f(z)$ for $z \in \mathcal{H}$. So f has 'Fourier expansion at infinity':

$$f(z) = \sum_{n=-\infty}^{\infty} a_n q_M^n \text{ where } q_M = e^{2\pi i z/M}.$$

We say that f is *holomorphic at infinity* if $a_n = 0$ for $n < 0$ and f *vanishes at infinity* if $a_n = 0$ for $n \leq 0$. Let $\sigma \in SL_2(\mathbb{Z})$. Then $\sigma^{-1}\Gamma\sigma$ also has finite index and

$$(f|\sigma)|\gamma = f|\sigma \text{ for all } \gamma \in \sigma^{-1}\Gamma\sigma.$$

So for any $\sigma \in SL_2(\mathbb{Z})$, $f|\sigma$ also has Fourier expansion at infinity. We say that f is *holomorphic at the cusps* if $f|\sigma$ is holomorphic at infinity for all $\sigma \in SL_2(\mathbb{Z})$. We say that f *vanishes at the cusps* if $f|\sigma$ vanishes at infinity for all $\sigma \in SL_2(\mathbb{Z})$.

Let $N \geq 1$ be an integer and ϵ a Dirichlet character $\pmod N$. A *modular form on* $\Gamma_0(N)$ *of type* (k, ϵ) is a holomorphic function f on $\mathcal{H}$ such that

$$(i)\ f\left|\begin{pmatrix} a & b \\ c & d \end{pmatrix}\right. = \epsilon(d) f \text{ for all } \begin{pmatrix} a & b \\ c & d \end{pmatrix} \in \Gamma_0(N)$$

and

$$(ii)\ f \text{ is holomorphic at the cusps.}$$

The Fourier expansion of such a form is

$$f(z) = \sum_{n=0}^{\infty} a_n q^n \text{ where } q = e^{2\pi i z}.$$

The integer k is called the *weight* of f. Such a modular form is called a *cusp form* if it vanishes at the cusps. The modular forms on $\Gamma_0(N)$ of type (k, ϵ) form a complex vector space denoted as $M_k(\Gamma_0(N), \epsilon)$ and this has a subspace $S_k(\Gamma_0(N), \epsilon)$ consisting of cusp forms. This subspace has a canonical complement:

$$M_k(\Gamma_0(N), \epsilon) = \mathcal{E}_k(\Gamma_0(N), \epsilon) \oplus S_k(\Gamma_0(N), \epsilon).$$

The space $\mathcal{E}_k(\Gamma_0(N), \epsilon)$ is called the space of Eisenstein series. These spaces are finite dimensional. One can define an *inner product* known as Petersson product on $S_k(\Gamma_0(N), \epsilon)$ by

$$\langle f, g \rangle = \int_{\mathcal{H}/\Gamma_0(N)} f(z)\overline{g(z)} y^k \frac{dxdy}{y^2}.$$

When ϵ is the *trivial character*, we omit ϵ in all the notations where it appears.

Hecke Operators. Let p denote a prime number and $f(z) = \sum_{n=0}^{\infty} a_n q^n$ be a modular form on $\Gamma_0(N)$ of type (k, ϵ). Define *Hecke operators* T_p and U_p as follows.

$$f|T_p = \sum_{n=0}^{\infty} a_n q^n + \epsilon(p) p^{k-1} \sum_{n=0}^{\infty} a_n q^{np} \text{ if } p \nmid N;$$

$$f|U_p = \sum_{n=0}^{\infty} a_{np} q^n \text{ if } p|N.$$

It is known that $f|T_p$ and $f|U_p$ are also modular forms of type (k, ϵ) and they are cusp forms if f is a cusp form.

Theorem 1.6.1 *The T_p's are commuting linear transformations of $S_k(\Gamma_0(N), \epsilon)$ and this space can be decomposed as a direct sum of eigenspaces.*

Let $f \in S_k(\Gamma_0(N), \epsilon)$. We say that f is an eigenform if f is an eigenfunction for all the Hecke operators T_p's and U_p's. If the Fourier coefficient a_1 of f equals 1, f is called *normalized*. Two eigenforms are said to be *equivalent* if they are in the same eigenspace in $S_k(\Gamma_0(N), \epsilon)$ under the action of the Hecke operators.

Old forms and newforms. Suppose $N'|N$ and ϵ is a Dirichlet character modN'. Suppose $f(z) \in S_k(\Gamma_0(N'), \epsilon)$ and $dN'|N$. Then

$$f(dz) \in S_k(\Gamma_0(N), \epsilon).$$

The forms on $\Gamma_0(N)$ which may be obtained in this way from divisors N' of N, $N' \neq N$ span a subspace $S_k^{old}(\Gamma_0(N), \epsilon)$, called the space of *oldforms*. Its orthogonal complement under the Petersson inner product is denoted as $S_k^{new}(\Gamma_0(N), \epsilon)$ and the eigenforms in this space are called *newforms*. Thus we have

$$S_k(\Gamma_0(N), \epsilon) = S_k^{old}(\Gamma_0(N), \epsilon) \oplus S_k^{new}(\Gamma_0(N), \epsilon).$$

Theorem 1.6.2 *If f is a newform then its equivalence class is one-dimensional.*

A new form has the Fourier expansion, say

$$f = q + \sum_{n \geq 2} c_n q^n.$$

Let $K = \mathbb{Q}(c_2, c_3, \ldots)$ be the field obtained by adjoining to $\mathbb{Q}$ the Fourier coefficients of f. This turns out to be a *totally real* number field. The Fourier coefficients are algebraic integers of K. We say that a newform is *rational* if the field $K = \mathbb{Q}$, in other words, if all the Fourier coefficients are in $\mathbb{Z}$. Let L be the algebraic closure of K. For any $\sigma \in Gal(L/\mathbb{Q})$, let

$$\sigma(f) = q + \sum_{n \geq 2} \sigma(c_n) q^n.$$

Then $\sigma(f)$ is also a newform. Usually we identify a newform with all its conjugates. For a given level N, the number of newforms (up to conjugacy or not) is finite. There are formulas for the number of newforms. Using them we get [24, Theorem 1]

Corollary 1.6.3 *There are no newforms for levels*

$$1, 2, 3, 4, 5, 6, 7, 8, 9, 10, 12, 13, 16, 18, 22, 25, 28, 60.$$

For all other levels there are newforms.

Rational Curves. A *plane algebraic curve* over a field $\mathbb{F}$ is the set of all points $(x, y) \in \overline{\mathbb{F}} \times \overline{\mathbb{F}}$ satisfying an equation of the form

$$f(x, y) = 0$$

where $f(X, Y) \in \mathbb{F}[X, Y]$. If f is irreducible, then the curve is called irreducible. An irreducible plane algebraic curve is called *rational* if there exist rational functions $r(t), s(t) \in \mathbb{F}(t)$, of which at least one is not constant, such that

$$f(r(t), s(t)) = 0$$

identically in t. For example, a circle is a rational curve. A compact Riemann surface of genus 0 is a rational curve over $\mathbb{C}$. If $\mathbb{F}$ is an algebraic number field and C is a curve over $\mathbb{F}$ of genus zero, with at least one $\mathbb{F}$-rational point, then C has infinitely many $\mathbb{F}$-rational points. The set of $\mathbb{F}$- rational points is denoted as $C(\mathbb{F})$. Two curves are said to be isomorphic if there is a birational transformation between them.

Elliptic curves over arbitrary fields. When $f(x, y)$ is of the form

$$y^2 + a_1xy + a_3y = x^3 + a_2x^2 + a_4x + a_6$$

where $a_i \in \mathbb{F}$ with a non-vanishing discriminant Δ (defined below), then we call the curve associated to this equation as an *elliptic curve* denoted by E and the above equation of f is known as the Weierstrass equation. If the characteristic of $\mathbb{F}$ is not 2, this equation can be put in the form

$$y^2 = 4x^3 + b_2x^2 + b_4x + b_6 \tag{1.17}$$

where

$$b_2 = a_1^2 + 4a_2;\ b_4 = 2a_4 + a_1a_3\ ;\ b_6 = a_3^2 + 4a_6.$$

Let us define the quantities:

$$b_8 = a_1^2a_6 + 4a_2a_6 - a_1a_3a_4 + a_2a_3^2 - a_4^2;$$

$$c_4 = b_2^2 - 24b_4\ ;\ c_6 = b_2^3 + 36b_2b_4 - 216b_6;$$

$$\Delta = -b_2^2b_8 - 8b_4^3 - 27b_6^2 + 9b_2b_4b_6\ ;\ j_E = c_4^3/\Delta.$$

Δ is called the *discriminant* of E and j_E is called the *j-invariant* of E. If the characteristic of $\mathbb{F}$ denoted as char($\mathbb{F}$) $\neq 2, 3$ the equation can be further simplified as

$$y^2 = x^3 - 27c_4x - 54c_6$$

known as Weierstrass normal form. If we write $y^2 = x^3 + ax + b$, then

$$\Delta = -16(4a^3 + 27b^2);\ j_E = 1728(4a)^3/\Delta.$$

Two elliptic curves are isomorphic over $\bar{\mathbb{F}}$ if and only if their j—invariants are equal. Let us specialize $\mathbb{F} = \mathbb{Q}$. Let E be an elliptic curve. Of all the Weierstrass equations for E, choose the one such that $|\Delta|$ is minimal and call that the *discriminant* of E denoted as Δ_E. Let the equation of E be given by

$$y^2 = x^3 + ax + b.$$

We say that E has *good reduction* at a prime p if $p \nmid \Delta_E$. The reason being when we reduce $E \mod p$ we will still get an elliptic curve over $\mathbb{F}_p$. If $p|\Delta$ and $p \nmid 2a$, we say that E has *multiplicative reduction* at p. Otherwise, we say that E has *additive reduction* at p. The *conductor* of E is defined by

$$N = \prod_{p|\Delta_E} p^{\delta_p}$$

where $\delta_p = 1$ if E has multiplicative reduction at p and δ_p is an integer ≥ 2 when E has additive reduction at p. Further it is known that $\delta_p = 2$ if $p \neq 2, 3$ and E has additive reduction at p. E is called *semistable* if its conductor is square free, i.e., if E has either good or multiplicative reduction everywhere.

Minimal Models. Let K be a field with discrete valuation ν. Let E be an elliptic curve over K. A *minimal normal form* for E is a normal form with all the a_j's in the valuation ring $\mathcal{O}_K$ of K such that $\nu(\Delta_E)$ is minimal among all such equations. The following proposition helps in finding such a minimal model.

Proposition 1.6.4 *Suppose that the residual characteristic of $K \neq 2, 3$. A minimal model satisfies:*

$$\nu(\Delta) + \min\{\nu(j), 0\} < 12 + 12\nu(2) + 6\nu(3).$$

For all K, the equation is minimal if

$$\nu(\Delta) < 12 \text{ or if } \nu(c_4) < 4 \text{ or if } \nu(c_6) < 6.$$

Conversely, any model that satisfies this inequality is minimal at ν.

If the residual characteristic $= 2$ or 3, there is an algorithm due to Tate which determines the minimal model at ν, in a finite number of steps.

The Frey curves $E_{A,B,C}$. Suppose $A, B, C \in \mathbb{Z}$ be such that $A + B = C$ and $\gcd(A, B, C) = 1$. Form the curve

$$E_{A,B,C} : Y^2 = X(X - A)(X + B)$$

which has j-invariant

$$j_{A,B,C} = 2^8 \frac{(A^2 + B^2 + AB)^3}{(ABC)^2}.$$

Notice that $A^2 + B^2 + AB$ and ABC are coprime. The discriminant of the equation $E_{A,B,C}$ is $(ABC)^2$. So it is minimal outside of 2 and 3. Suppose $3 \nmid ABC$. Then the discriminant is minimal at 3. Suppose now $3|ABC$. Then

$$c_4 = 16(A^2 + B^2 + AB)$$

is not divisible by 9. Hence by the Proposition 1.6.4 the equation is minimal at 3. Now suppose that

$$A \equiv -1 \pmod 4;\ B \equiv 0 \pmod{32}.$$

Make the following change of variables

$$x = 4X,\ y = 8Y + 4X.$$

Then the equation for $E_{A,B,C}$ becomes

$$Y^2 + XY = X^3 + eX^2 + dX$$

where

$$e = \frac{B + 1 - A}{4};\ d = -\frac{AB}{16}$$

having discriminant

$$\Delta = d^2(B - A)^2 - 64d^3 = d^2(A + B)^2$$

and j-invariant

$$j = \frac{(A^2 + B^2 + AB)^3}{\Delta}.$$

Now

$$\nu_2(\Delta) = 2\nu_2(B) - 8;\ c_4 = (1 + 4e)^2 - 48d.$$

So c_4 is odd and hence the equation is minimal at 2. The discriminant of $E_{A,B,C}$ is

$$\frac{(ABC)^2}{2^8}.$$

Thus $E_{A,B,C}$ is semistable outside of 2 and if $A \equiv -1 \pmod 4$; $B \equiv 0 \pmod{32}$ then $E_{A,B,C}$ is semistable at 2 also.

The group law. The set $E(\mathbb{F})$ of all points (x, y) on E with $x, y \in \mathbb{F}$ can be given the structure of an additive abelian group when it is non-empty. We view the curve in $\mathbb{P}^2$ (projective space of dimension 2) as

$$Y^2Z + a_1XYZ + a_3YZ^2 = X^3 + a_2X^2Z + a_4XZ^2 + a_6Z^3.$$

Then it has $(0, 1, 0)$ as a solution. We will call this solution as the *point at infinity* denoted by $\mathcal{O}$. For any elliptic curve E, we denote the n-torsion subgroup $E[n]$ to be the set of points on E of order dividing n:

$$E[n] = \{P \in E : nP = \mathcal{O}\}.$$

In fact, $E[n] \simeq (\mathbb{Z}/n\mathbb{Z}) \oplus (\mathbb{Z}/n\mathbb{Z})$. As an example, if we want to get points of order 2 in $E(\mathbb{F})$ when E is given by (1.17), we solve the cubic

$$4x^3 + b_2x^2 + b_4x + b_6 = 0.$$

We get three roots e_1, e_2, e_3 so that $(e_1, 0), (e_2, 0), (e_3, 0)$ are the three points of order 2. A point is a *singular point* of the curve $f(x, y) = 0$ if

$$\frac{\partial f}{\partial x}(P) = \frac{\partial f}{\partial y}(P) = 0.$$

We denote by E_{ns} and $E_{ns}(\mathbb{F})$, the set of non-singular points of E and $E(\mathbb{F})$, respectively.

Theorem 1.6.5 *Let E be a curve given by Weierstrass equation with discriminant $\Delta = 0$. (So E has a singular point). Then under the composition law of points on E, $E_{ns}(\bar{\mathbb{F}})$ is an abelian group. Further*

$$E_{ns}(\bar{F}) \simeq \begin{cases} \bar{\mathbb{F}}^{\times} & \textit{if } c_4 \neq 0; \\ \bar{\mathbb{F}}^{+} & \textit{if } c_4 = 0. \end{cases}$$

Elliptic curves over number fields. Poincaré conjectured that if $\mathbb{F}$ is an algebraic number field, $E(\mathbb{F})$ is finitely generated as a $\mathbb{Z}$-module. This was proved by Mordell in the case $\mathbb{F} = \mathbb{Q}$ and in the general case of a number field or a function field over a finite field by A. Weil.

Theorem 1.6.6 (Mordell-Weil) *If $\mathbb{F}$ is a global field, then*

$$E(\mathbb{F}) \simeq \mathbb{Z}^{r_{\mathbb{F}}} \oplus E(\mathbb{F})_{tor}.$$

$r_{\mathbb{F}}$ is called the *rank of E over* $\mathbb{F}$. When $\mathbb{F} = \mathbb{Q}$, $E(\mathbb{Q})_{tor}$ has a very nice description.

Theorem 1.6.7 (Lutz-Nagell) *Let $E/\mathbb{Q}$ be an elliptic curve with Weierstrass equation*

$$y^2 = x^3 + ax + b, a, b \in \mathbb{Z}.$$

If $P \in E(\mathbb{Q})_{tor}$, $P \neq 0$, then $x(P), y(P) \in \mathbb{Z}$. Also, either

$$2P = 0 \textit{ or } y(P)^2 | (4a^3 + 27b^2).$$

Example. Let $E : y^2 = x^3 - 2$ over $\mathbb{Q}$. The discriminant is 108. Considering the equation mod 5 and mod 7 we find that

$$E(\mathbb{F}_5) = \{(1, \pm 2), (2, \pm 1), (3, 0), \mathcal{O}\}$$

so that $|E(\mathbb{F}_5)| = 6$;

$$E(\mathbb{F}_7) = \{(3, \pm 2), (5, \pm 2), (6, \pm 2), \mathcal{O}\}$$

so that $|E(\mathbb{F}_7)| = 7$. Since $E(\mathbb{Q})_{tor} \subset \mathbb{Z} \times \mathbb{Z}$, $E(\mathbb{Q})$ injects into $E(\mathbb{F}_5)$ and $E(\mathbb{F}_7)$. Since they have coprime orders, $E(\mathbb{Q})_{tor} = 1$. Hence $(3, 5)$ is a point of infinite order.

Theorem 1.6.8 *(Mazur) If $E/\mathbb{Q}$ is an elliptic curve, then $(E/\mathbb{Q})_{tor}$ is one of the 15 groups:*

$$\mathbb{Z}/n\mathbb{Z},\ 1 \le n \le 10 \text{ or } n = 12;$$

$$\mathbb{Z}/2\mathbb{Z} \oplus \mathbb{Z}/2n\mathbb{Z},\ 1 \le n \le 4$$

and each of these groups occurs.

There is no effective procedure known to compute the rank of an elliptic curve. The Birch and Swinnerton-Dyer conjectures are an attempt to give an effective procedure.

The L-function of an elliptic curve. Let E be an elliptic curve over $\mathbb{Q}$. Define

$$a_p(E) = a_p = p + 1 - |E_{ns}(\mathbb{F}_p)|.$$

The L-function of E is defined as

$$L(E, s) = \prod_{p|N} \left(1 - \frac{a_p}{p^s}\right)^{-1} \prod_{p\nmid N} \left(1 - \frac{a_p}{p^s} + \frac{1}{p^{2s-1}}\right)^{-1}$$

where N is the conductor of E. The Birch and Swinnerton-Dyer conjecture is that $L(E, s)$ extends to an entire function and

$$\mathrm{ord}_{s=1} L(E, s) = rank_{\mathbb{Z}}(E(\mathbb{Q})).$$

The conjecture has been proved in a few cases, for instance when $r_{\mathbb{Q}} = 0$ or 1.

Isogenies. Let E_1 and E_2 be two elliptic curves over a field F. We say two elliptic curves E_1 and E_2 are *isogenous* if there is an epimorphism $\phi : E_1(\bar{F}) \to E_2(\bar{F})$ with finite kernel. If ϕ is defined over F then we say that E_1 and E_2 are isogeneous over F. An *endomorphism* of an elliptic curve E is an isogeny from E to itself. The set of all endomorphisms of E forms a ring known as *endomorphism ring* of E, denoted as $End(E)$. For example, *multiply by n map* which sends a point $P \in E$ to $nP \in E$ is an endomorphism. In particular, every endomorphism ring contains a subring isomorphic to $\mathbb{Z}$. Consider the elliptic curve $E : y^2 = x^3 + x$. This has an endomorphism

$$\phi(x, y) = (-x, iy)$$

with ϕ^4 the identity map. But (x, y) is not generally a point of any particular order. So ϕ is not a multiply by n map. Thus $End(E) \neq \mathbb{Z}$. Such curves, i.e., elliptic curves E with $End(E) \supsetneq \mathbb{Z}$ are called elliptic curves with *complex multiplication.*

Theorem 1.6.9 *If E_1 and E_2 are isogenous elliptic curves over $\mathbb{Q}$, then $L(E_1, s) = L(E_2, s)$.*

In 1967, Tate conjectured that the converse holds, i.e., if two elliptic functions have the same L-function, then they are isogenous. This was proved by Faltings. Having breezed through some classical details, we now plunge into a few modern developments. For details and clarifications, the reader is directed to [24].

Rational newforms and Elliptic curves. We state one of the important theorems in this theory, main contributors being Shimura, Wiles, Taylor, Breuil, Conrad, Diamond and Taylor [24, Theorem 2].

Theorem 1.6.10 (The Modularity theorem for elliptic curves) *Let $N \geq 1$ be an integer. There is a one-to-one correspondence $f \to E_f$ between rational new forms of level N and isogeny classes of elliptic curves E defined over $\mathbb{Q}$ and of conductor equal to N. Under this correspondence, for all primes $\ell \nmid N$, we have $c_\ell = a_\ell(E_f)$, where c_ℓ is the ℓth Fourier coefficient of f and $a_\ell(E_f) = \ell + 1 - |E_f(\mathbb{F}_\ell)|$ where $\mathbb{F}_\ell$ denotes the finite field with order ℓ.*

Let E be an elliptic curve over $\mathbb{Q}$ of conductor N. Let f be a newform of level N' and let K be the associated number field. We say that E *arises modulo p from* f and write $E \sim_p f$, if there exists a prime ideal $\mathfrak{p}$ of K above p such that $c_\ell \equiv a_\ell(E) \pmod{\mathfrak{p}}$ for all but finitely many prime numbers ℓ. For example, if $E = E_f$ is the elliptic curve of conductor N' corresponding to a rational newform f, then $c_\ell = a_\ell(E)$ for $\ell \nmid N'$ so that $E \sim_p f$ for all primes p. On the other hand, if E is an elliptic curve of conductor N such that $E \sim_p f$ with f a rational newform of level N' then by modularity theorem we know that f corresponds to an elliptic curve $F = E_f$ defined over $\mathbb{Q}$ of conductor N'. The following result gives some properties.

Theorem 1.6.11 *Let E be an elliptic curve over $\mathbb{Q}$ of conductor N. Let f be a newform of level N' and let K be the associated number field. Assume that $E \sim_p f$. There exists a prime ideal $\mathfrak{p}$ of K above p such that for all prime numbers ℓ we have*

(i) If $\ell \nmid pNN'$, then $a_\ell(E) \equiv c_\ell \pmod{\mathfrak{p}}$.
(ii) If $\ell || N$ but $\ell \nmid pN'$ then $c_\ell \equiv \pm(\ell + 1) \pmod{\mathfrak{p}}$.

This is [24, Proposition 5.1]. Let E be an elliptic curve over $\mathbb{Q}$ which is a minimal model with discriminant Δ_{min}. Let the conductor of E be N. Then $N | \Delta_{min}$ and N and Δ_{min} have the same prime divisors, the primes of bad reduction. Let us define

$$N_p = N \Bigg/ \prod_{\substack{q || N \\ p | \nu_q(\Delta_{min})}} q.$$

We now state a simplified version of Ribet's level-lowering theorem [24, Theorem 3].

Theorem 1.6.12 *Let E be an elliptic curve defined over $\mathbb{Q}$ and let $p \geq 5$ be a prime number. Assume that there does not exist a p-isogeny (i.e., of degree p) defined over $\mathbb{Q}$ from E to some other elliptic curve. Then there exists a newform f of level N_p such that $E \sim_p f$.*

Example. Let E be the elliptic curve with minimal Weierstrass equation

$$y^2 = x^3 - x^2 - 77x + 330$$

(refer 132B1 in [27]). Then

$$\Delta_{min} = 2^4 \cdot 3^{10} \cdot 11; N = 2^2 \cdot 3 \cdot 11.$$

The only isogeny that the curve has is a 2-isogeny. Applying Theorem 1.6.12 with $p = 5$ we find, $N_p = 2^2 \cdot 11 = 44$, so that there exists a newform of level 44 such that $E \sim_5 f$. It is known that there is a single newform at level 44 which is necessarily rational. Cremona's table shows that it corresponds to the elliptic curve $F = 44A1$ with equation

$$y^2 = x^3 + x^2 + 3x - 1$$

so that $E \sim_5 F$. There are several results to decide when E will not have any p-isogeny over $\mathbb{Q}$ which is a pre-condition to apply Theorem 1.6.12. We state below a few [24, Theorems 4 and 5].

Theorem 1.6.13 *Let E be an elliptic curve defined over $\mathbb{Q}$ of conductor N. Then E does not have any p-isogeny if at least one of the following conditions hold:*

(a) $p \geq 17$ and $j(E) \notin \mathbb{Z}[1/2]$.
(b) $p \geq 11$ and N is square free.
(c) $p \geq 5$, N is square free, and $4 \mid |E(\mathbb{Q})_{tor}|$.
(d) If $\nu_2(N) = 3, 5$ or 7.

For instance, suppose E be such that $\Delta_{min} = 11^{11}$ and $N = 11$. Then $N_p = 1$. Then by (b) above, E does not have any 11-isogeny. So by Theorem 1.6.12, $E \sim_{11} f$ for a newform f of level 1. But there is no such newform. Thus there is no such elliptic curve.

Some applications. Soon after the resolution of Fermat's Last Theorem, many other Diophantine equations were solved using the above ideas. We shall restrict here to two results. The first result is [29, Lemma 13] and it is inspired by the work of Ribet.

Lemma 1.6.14 *Let $\ell \geq 5$. Let a, b, c be non-zero integers such that either $P(abc) \leq 3$ or a, b, c are composed of only 2 and 5. Then the equation*

$$ax^\ell - by^\ell = cz^\ell$$

with $\gcd(ax^\ell, by^\ell, cz^\ell) = 1$ and $\nu_2(by^\ell) \geq 4$ has no solution in non-zero integers x, y, z.

Proof Suppose x, y, z are non-zero integers satisfying the given equation. Put $A = ax^\ell$, $B = -by^\ell$ and $C = cz^\ell$. Thus

$$A + B = C.$$

Note that A and C are odd, $\gcd(A, B) = 1$ and $16|B$. We may assume that

$$A \equiv -1 \pmod 4$$

by possibly multiplying the given equation by -1. Let the Frey elliptic curve be E:

$$Y^2 = X(X - A)(X + B).$$

The conductor N_E of this curve is given by

$$N_E = 2^t R(ABC)$$

where according to computations made by Diamond and Kramer using Tate's algorithm,

$$t = \begin{cases} 0 \text{ if } \nu_2(by^\ell) = 4 \\ 1 \text{ if } \nu_2(by^\ell) > 4 \end{cases}$$

and

$$R(ABC) = \prod_{\substack{p|ABC \\ p-\text{odd prime}}} p.$$

Thus N_E is square free. Further the minimal discriminant Δ_E is given by

$$\Delta_E = 2^u (ABC)^2$$

where u is an integer. So if $p|\Delta_E$ we have

$$\nu_p(\Delta_E) \equiv 0 \pmod \ell$$

except when $p = 2, 3$ if $P(abc) \leq 3$ and except $p = 2, 5$ if a, b and c are composed of only 2 and 5. Hence by Theorem 1.6.12, $E \sim_\ell f$ for some newform f of level N_ℓ, i.e., there exists a newform of level 1, 2, 3 or 6 if $P(abc) \leq 3$ or a newform of level 1, 2, 5 or 10 if a, b, c are composed of only 2 and 5. We see from Corollary 1.6.3 that there are no newforms of level 1, 2, 3, 5, 6, 10. This is a contradiction. Hence the given equation has no solution in non-zero integers x, y, z. □

Note. The above theorem includes Fermat's Last Theorem.

Lemma 1.6.15 *Let $\ell \geq 3$ and α an integer with $1 \leq \alpha < \ell$. Then the equation*

$$x^\ell + y^\ell = 2^\alpha z^\ell$$

in non-zero relatively prime integers x, y, z has no solution for $\alpha > 1$ and for $\alpha = 1$ the equation has only the trivial solution for which $xyz = \pm 1$.

The above assertion for $\alpha > 1$ is due to Ribet [30] and the case $\alpha = 1$ is due to Darmon and Merel [31]. Here we shall prove the theorem for $\alpha = 1$ which exhibits more complication [24, Theorems 6 and 7].

Lemma 1.6.16 *Let $\ell \geq 5$ be a prime. The equation*

$$x^\ell + 2y^\ell = z^\ell, \gcd(a, b, c) = 1$$

has no solution with $xyz \neq 0$ except when $xyz = \pm 1$.

Proof The theorem for $\ell \leq 31$ was achieved by Dénes [32] using classical algebraic number theory methods. So we shall assume that $\ell > 31$, to simplify our work. Assume that (x, y, z) is a non-trivial primitive solution to the given equation. Note that x and z are odd. By multiplying by -1, if necessary, we may assume that $x^\ell \equiv -1 \pmod 4$. We associate with this the Frey curve

$$E : Y^2 = X(X - x^\ell)(X + 2y^\ell).$$

We compute

$$c_4 = 16(z^{2\ell} - 2(xy)^\ell), \Delta(E) = 2^6(xyz)^{2\ell}, j(E) = \frac{2^6(z^{2\ell} - 2(xy)^\ell)^3}{(xyz)^{2\ell}}.$$

Note that $\Delta(E)$ is not necessarily minimal. Using Tate's algorithm, the minimal discriminant can be computed as

$$\Delta_{min} = \begin{cases} 2^6(xyz)^{2\ell} \text{ if } 16 \nmid 2y^\ell \\ (xyz)^{2\ell}/2^6 \text{ if } 16 | 2y^\ell. \end{cases}$$

Further

$$N_E = \begin{cases} 8 \prod_{\substack{p|xyz \\ p\neq 2}} p \text{ if } y \text{ is even} \\ 32 \prod_{\substack{p|xyz \\ p\neq 2}} p \text{ if } y \text{ is odd.} \end{cases} \tag{1.18}$$

Since $\nu_2(N_E) = 3$ or 5 by Theorem 1.6.13 (*d*), there is no p-isogeny. Hence by Theorem 1.6.12, there exists a newform of level N_ℓ such that $E \sim_\ell f$. Now

$$N_\ell = 8 \text{ or } 32 \text{ according as } y \text{ is even or odd.}$$

There is no newform at level 8. Hence $N_\ell = 32$ and y is odd. There are newforms at level $N_\ell = 32$. It is known that there is a single newform, so necessarily rational and using Cremona [27], it corresponds to the elliptic curve F (ref: 32A2) given by

$$F : Y^2 = X(X+1)(X+2).$$

This is the Frey curve corresponding to the solution $(x, y, z) = (-1, 1, 1)$ and we have $E \sim_\ell F$. Putting $X_1 = X + 1$, we get a curve isomorphic to F, viz., $Y^2 = X_1^3 - X_1$. Note that this curve has complex multiplication by i, i.e., changing (X_1, Y) to $(-X_1, iY)$ gives back the same curve. Note that $\text{conductor}(F) = N' = 32$. We consider two cases.

Case 1. $\ell \equiv 1 \pmod 4$. In this case, ℓ splits in $\mathbb{Z}[i]$. It is known in this case (by the works of Haberstadt and Kraus) since $\ell > 31$ that

$$\text{conductor}(E) = \text{conductor}(F) = 32.$$

This means by (1.18)

$$\prod_{\substack{p \mid xyz \\ p \neq 2}} p = 1, \; y \text{ odd}.$$

Hence x, y and z are powers of 2. But since x, y and z are all odd, we get $(x, y, z) = \pm(-1, 1, 1)$.

Case 2. $\ell \equiv 3 \pmod 4$. In this case ℓ is inert in $\mathbb{Z}[i]$. The Frey curve E clearly has rational 2-torsion (in fact full 2-torsion) so that (by a result of Darmon and Merel), $j(E) \in \mathbb{Z}[1/\ell]$. Note that by (1.18), $\ell^2 \nmid N_E$ and $\ell \nmid N_\ell$. In this case, it was shown (by Siksek [24]) that in fact, $j(E) \in \mathbb{Z}$. But $j(E) = 64(z^{2\ell} - 2x^\ell y^\ell)^3/(xyz)^{2\ell}$ and since x, y and z are pairwise coprime, it follows that x, y and z cannot have any odd prime divisors, hence they are powers of 2, so we conclude that $(x, y, z) = \pm(-1, 1, 1)$. □

Exercise 1

(1) Show that $\omega(n) \ll \frac{\log n}{\log \log n}$. Compute the constant involved in $\ll$ explicitly.
(2) Show that $\prod_{p \leq n} p < 4^n$ using induction and the fact that $\binom{n}{k} < 2^{n-1}$.
(3) Show that each term in the brackets on the right-hand side of (1.11) has the value 0 or 1.

References

1. B. Rosser, L. Schoenfeld, Approximate formulas for some functions of prime numbers. Illinois J. Math. **6**, 64–94 (1962) [**1**]
2. P. Dusart, Autour de la fonction qui compte le nombrede nombres premiers. Ph.D. thesis, University de Limoges, 1998. [**1**]

3. A. Baker, *Transcendental Number Theory*, 2nd edn. (Cambridge University Press, 1979). **[1]**
4. T.N. Shorey, R. Tijdeman, *Exponential Diophantine Equations* (Cambridge University Press, 1986). **[1]**
5. N. Saradha, R. Thangadurai, *Pillars of Transcendental Number Theory* (Springer Nature Singapore Pte Ltd., 2020), 184 p. **[1]**
6. A. Baker, The theory of linear forms in logarithms, in *Transcendence Theory: Advances and Applications* (Academic Press, 1977), pp. 1–27. **[1]**
7. T.N. Shorey, Perfect powers in values of certain polynomials at integer points. Math. Proc. Cambridge Philos. Soc. **99**, 268–273 (1986). **[1]**
8. A. Baker, Bounds for the solutions of hyperelliptic equations. Proc. Camb. Phil. Soc. **65**, 439–444 (1969). **[1]**
9. A. Baker, Rational approximations to $\sqrt[3]{2}$ and other algebraic numbers. Q. J. Math. Oxford **15**(2), 375–383 (1964). **[1]**
10. J.H. Evertse, On the equation $ax^n - by^n = c$. Compositio Math. **47**, 289–315 (1982). **[1]**
11. I. Niven, H. Zuckerman, H.L. Montgomery, *An Introduction to the Theory of Numbers*, 5th edn. (Wiley, 2001), pp. 529. **[1]**
12. M. Bennett, Rational approximation to algebraic numbers of small height: the diophantine equation $|ax^\ell - by^\ell| = 1$, J. Reine Angew. Math. **535**, 1–49 (2001). **[1]**
13. K.F. Roth, On the gaps between square-free numbers. J. London Math. Soc. **26**, 263–268 (1951). **[1]**
14. H. Halberstam, K.F. Roth, On the gaps between consecutive k-free integers. J. Lond. Math. Soc. **26**, 268–273 (1951). **[1]**
15. P. Erdős, On the product of consecutive integers III. Indag. Math. **17**, 85–90 (1955). **[1,4,5]**
16. H. Halberstam, H.E. Richert, *Sieve Methods* (Academic Press,1974). **[1]**
17. P. Erdős, A theorem of Sylvester and Schur. J. Lond. Math Soc. **9**, 282–288 (1934). **[1,3]**
18. A. Knapp, *Elliptic Curves* (Princeton University Press, 1992), p. 272. **[1]**
19. S. Lang, Introduction to modular forms. Grundl. Math. Wiss. **222** (Springer, Berlin, New York, 1976). **[1]**
20. A. Ogg, Survey of modular functions of one variable, in *Modular Functions of One Variable I*, ed. W. Kuijk. Lecture Notes in Mathematics, vol. 320 (Springer), pp 1–36. **[1]**
21. A. Ogg, *Modular Forms and Dirichlet Series* (W.A. Benjamin, 1969). **[1]**
22. J. Silvermann, *The Arithmetic of Elliptic Curves* (Springer, 1986). **[1]**
23. J. Silvermann, *Advanced topics in the Arithmetic of Elliptic Curves*. Graduate Texts in Mathematics, vol. 106 (Springer, Berlin-New York, 1986). **[1]**
24. S. Siksek, *The Modular Approach to Diophantine Equations*. Lecture Notes-Home Page. **[1]**
25. G. Shimura, Complex multiplication, in *Modular Functions of One Variable I* ed. W. Kuijk. Lecture Notes in Mathematics, vol. 320 (Springer, 1972), pp. 37–56. **[1]**
26. A. Dujella, *Number Theory* (Skolska knjiga, Zagreb, 2021), 636 p. **[1]**
27. J. Cremona, *Algorithms for Modular Elliptic Curves* (Cambridge University Press, 1992). **[1]**
28. W. Stein, Modular forms database, http://modular.fas.harvard.edu/Tables/ **[1, 8]**
29. N. Saradha, T.N. Shorey, Almost perfect powers in arithmetic progression. Acta Arith. **138**, 363–388 (2001). **[1, 6, 7]**
30. K.A. Ribet, On the equation $a^p + 2^\alpha b^p + c^p = 0$. Acta Arith. **79**, 7–16 (1997). **[1]**
31. H. Darmon, L. Merel, Winding quotients and some variants of Fermat's Last Theorem. J. Reine Angew. Math. **490**, 81–100 (1997). **[1]**
32. P. Dénes, Über die Diophantische Gleichung $x^\ell + y^\ell = cz^\ell$. Acta Math. **88**, 241–251 (1952). **[1, 5]**

Chapter 2
Basic Ideas of Erdős and Selfridge

$E \leq s + \binom{r}{2}$

Let

$$S = \{n_1 < n_2 < \cdots < n_{|S|}\}$$

with $n_i = a + t_i d$, $1 \leq i \leq |S|$ be a set of $|S| \leq k$ integers belonging to $\{a, a + d, \ldots, a + (k-1)d\}$ with $\gcd(a, d) = 1$. Let $\ell \geq 2$ be an integer. This book deals with equations of the form

$$n_1 \cdots n_{|S|} = (a + t_1 d) \cdots (a + t_{|S|} d) = by^\ell, \; P(b) \leq k. \tag{2.1}$$

We assume that there exist positive integers a, d, k, b, y, ℓ and $t_1, \ldots, t_{|S|} \in [0, k-1]$ such that (2.1) is satisfied. Note that we may assume that either $\ell = 2$ or ℓ is an *odd* prime which we shall suppose from now onwards. In this chapter, we bring together some of the basic and salient features like the fundamental argument of Erdős (Sect. 2.1), counting functions (Sect. 2.2), graph theoretic arguments (Sect. 2.3) and distinctness of certain products under restrictions on d (Sect. 2.4). Invariably, these ideas show up while dealing with (2.1).

2.1 A Fundamental Argument of Erdős

We describe an argument of Erdős which is essential to the problems dealt with in this book. Let us write each n_i as

$$n_i = a + t_i d = u_i v_i$$

S. Natarajan, *Perfect Powers—An Ode to Erdős*, Infosys Science Foundation Series,
https://doi.org/10.1007/978-981-96-2599-4_2

where u_i is composed of all primes $\leq k$, v_i is composed of all primes $> k$ and $\gcd(u_i, v_i) = 1$. Note also that primes dividing n_i are coprime to d since $\gcd(n, d) = 1$. We like to point out here that this gcd condition will be used at innumerable places without any mention. We have

$$u_i = \prod_{\substack{p \leq k \\ p \nmid d}} p^{\nu_p(n_i)}; \quad v_i = \prod_{\substack{p > k \\ p \nmid d}} p^{\nu_p(n_i)}.$$

Suppose a prime p divides two integers n_i and n_j, $i < j$. Then p divides $n_j - n_i = (t_j - t_i)d$. Since p cannot divide d, it divides $t_j - t_i < k$. Thus all primes $> k$ can occur in at most one of the n_i's. This shows that

$$v_1, \ldots, v_{|S|} \textit{ are all pairwise coprime.}$$

In fact, if k were a prime, it will also occur in at most one term. Let us now consider the primes $p \leq k$ dividing the n_i's. There are $\pi_d(k)$ such primes. Pick an integer among the n_i's in which p occurs to the maximum. Let us denote this integer as $n^{(p)} = a + t^{(p)}d$, say. There are at most $\pi_d(k)$ such integers. Remove from the set S these $\leq \pi_d(k)$ integers to get a subset $S_1 \subset S$ with

$$|S_1| \geq |S| - \pi_d(k).$$

Note that p divides terms like $a + (t^{(p)} + p)d, a + (t^{(p)} + 2p)d, a + (t^{(p)} + 3p) d, \ldots, a + (t^{(p)} - p)d, a + (t^{(p)} - 2p)d, \ldots$. We need to take only those which are in $S \subseteq [a, a + (k - 1)d]$. Suppose p divides the term $n_i > n^{(p)}$. Then $p^{\nu_p(n_i)}$ divides both n_i and $n^{(p)}$ where it occurs to the maximum power. So $p^{\nu_p(n_i)}$ divides $(t_i - t^{(p)})d$ and hence $p^{\nu_p(n_i)}$ divides $t_i - t^{(p)}$ where $1 \leq t_i - t^{(p)} \leq k - t^{(p)} - 1$. If $n_i < n^{(p)}$, then $1 \leq t^{(p)} - t_i < t^{(p)}$. Thus

$$\prod_{n_i \in S_1} u_i = \prod_{\substack{p \leq k \\ n_i \in S_1}} p^{\nu_p(n_i)} \text{ divides } \prod_{n_i \in S_1} |t_i - t^{(p)}|.$$

But

$$\prod_{n_i \in S_1} |t_i - t^{(p)}| \text{ divides } (t^{(p)} - 1)!(k - t^{(p)} - 1)!$$

which gives

$$\prod_{n_i \in S_1} u_i \,\Big|\, (k - 1)! \tag{2.2}$$

Observe that the left-hand side product is not divisible by any prime $p|d$. Hence we get

$$\prod_{n_i \in S_1} u_i \,\Bigg|\, \Bigg((k - 1)! \prod_{\substack{p|d \\ p \leq (k-1)}} p^{-\nu_p((k-1)!)}\Bigg). \tag{2.3}$$

In particular,

$$\prod_{n_i \in S_1} u_i \le \left((k-1)! \prod_{\substack{p|d \\ p \le (k-1)}} p^{-\nu_p((k-1)!)}\right). \tag{2.4}$$

Equation (2.3) can be further improved. For instance, suppose we know the order of a prime q dividing the left-hand side product, say α_q. Then

$$\prod_{n_i \in S_1} u_i q^{-\alpha_q} \Bigg| \left((k-1)! \prod_{\substack{p|d \\ p \le (k-1)}} p^{-\nu_p((k-1)!)} q^{-\nu_q (k-1)!}\right),$$

i.e.,

$$\prod_{n_i \in S_1} u_i \Bigg| \left((k-1)! \prod_{\substack{p|d \\ p \le (k-1)}} p^{-\nu_p((k-1)!)} q^{\alpha_q - \nu_q (k-1)!}\right). \tag{2.5}$$

This can be done for more than one prime q. We shall henceforth call the argument in this section as *the fundamental argument of Erdős.*

2.2 Counting Functions

Two representations of n_i's.

Let us consider $n_i = a + t_i d = u_i v_i$ once again. Here $\gcd(u_i, v_i) = 1$, v_i's are pairwise coprime and $P(u_i) \le k$. Thus every prime dividing v_i appears to an ℓ th power by (2.1). So

$$v_i = y_i^\ell.$$

We can always write u_i as

$$u_i = a_i e_i^\ell \text{ with } a_i - \ell\text{th power free, } P(a_i) \le k.$$

Suppose k were a prime, it will divide only one term, say $a + t_i d$. Then $P(a_i) \le k$ and $P(a_j) < k$ for $j \ne i$. Also if $P(b) < k$, then $P(a_i) < k$ for $1 \le i \le |S|$. Let

$$n_i = a + t_i d = a_i x_i^\ell,\ P(a_i) \le k,\ a_i - \ell\text{th power free} \tag{2.6}$$

and

$$n_i = a + t_i d = A_i X_i^\ell,\ P(A_i) \le k,\ \gcd(A_i, X_i) = 1. \tag{2.7}$$

Put

$$T(a) = \{a_0, \ldots, a_{|S|}\},\ T(A) = \{A_0, \ldots, A_{|S|}\}.$$

Note that if $X_i > 1$, then it is composed of only primes $> k$. Since these primes occur only in one term of the product we get $\gcd(X_i, X_j) = 1$. The case of squares, viz., $\ell = 2$ is of special interest. In this case we see that a_i's are square free. So any prime divides a_i if and only if it appears to an odd power in $n + t_i d$. Let $p^\alpha \le k < p^{\alpha+1}$. By inclusion-exclusion principle, the number of terms in S divisible by odd power of p is at most

$$\left\lceil \frac{k}{p} \right\rceil - \left\lfloor \frac{k}{p^2} \right\rfloor + \left\lceil \frac{k}{p^3} \right\rceil - \left\lfloor \frac{k}{p^4} \right\rfloor + \cdots$$
$$\le \left(\frac{k}{p} - \frac{k}{p^2}\right)\left(1 + \frac{1}{p^2} + \frac{1}{p^4} + \cdots\right) + 2\alpha$$
$$\le \frac{k}{p+1} + \frac{2\log k}{\log p}.$$

Hence

$$\nu_p\left(\prod_{n_i \in S} a_i\right) \le \frac{k}{p+1} + \frac{2\log k}{\log p}. \tag{2.8}$$

On many occasions we need to count number of a_i's or A_i's with some given properties. For instance, we know that $P(a_i) \le k$, i.e., a_i's are composed of primes from the set

$$L = \{p_1, \ldots, p_{\pi(k)}\}.$$

Now we ask how many a_i's are composed of only the primes, say $p_1, \ldots p_m$ from L. Let $T(a)(k, m)$ be the set of such a_i's and let $f(k, m) = |T(a)(k, m)|$. Let ϵ_p be the number of a_i's divisible by p. Then

$$\epsilon_p \le \left\lceil \frac{k}{p} \right\rceil \text{ if } p \nmid d \text{ and } p \nmid k$$

and

$$\epsilon_p \le \left\lfloor \frac{k}{p} \right\rfloor \text{ if } p \nmid d \text{ and } p \mid k.$$

We put

$$\epsilon_p = 0 \text{ if } p|d.$$

Hence

$$f(k, m) \ge |T(a)| - \sum_{p \in L \setminus \{p_1, \ldots, p_m\}} \epsilon_p := f_0(k, m). \tag{2.9}$$

Let us illustrate the use of this with an example. Let

$$S = \{n, n+1, \ldots, n+k-1\} \text{ and } \ell = 2.$$

Since a_i's are square free we get

$$f(k,m) \le 2^m. \tag{2.10}$$

We now fix $k = 9$ and $m = 2$. Assume a_i's are all distinct. So $|T(a)| = 9$, all a_i's are composed of 2, 3, 5, 7 and we want to find the number of a_i's composed of only 2 and 3. By (2.9), it is at least

$$9 - 2 - 2 = 5$$

which contradicts (2.10) with $m = 2$. So we conclude that all a_i's are *not* distinct. This counting function is useful for small values of k. As k increases $f_0(k,m)$ becomes ≤ 0 hence useless. For moderately large values of k, another counting function was proposed in [1, p. 298]. Let us write the sequence of primes as

$$p_1 < \cdots < p_{m_1} < p_{m_1+1} < \cdots < p_{m_1+m_2} < \cdots.$$

We now look for the number of a_i's which are composed of primes $\le p_{m_1+m_2}$, but divisible at most to the *first* power by only *one* prime from $\{p_{m_1+1}, \ldots, p_{m_1+m_2}\}$. Let $T(a)(k, m_1, m_2)$ be the set of such a_i's and $F(k, m_1, m_2) = |T(a)(k, m_1, m_2)|$. We want to get a lower bound for $F(k, m_1, m_2)$. Note that for each prime $p > p_{m_1+m_2}$ and $\le k$ we omit at most ϵ_p number of a_i's. Also we omit at most $\left\lceil \frac{k}{p_i p_j} \right\rceil$ number of a_i's for $m_1 + 1 \le i \le j \le m_1 + m_2$. Thus

$$\begin{aligned} F(k, m_1, m_2) &\ge |T(a)| - \sum_{p_{m_1+m_2} < p \le k} \epsilon_p - \sum_{1 \le i \le j \le m_2} \left\lceil \frac{k}{p_{m_1+i} p_{m_1+j}} \right\rceil \\ &\ge |T(a)| - \sum_{p_{m_1+m_2} < p \le k} \epsilon_p - \left(\frac{k}{2} \left(\sum_{i=1}^{m_2} \frac{1}{p_{m_1+i}^2} + \left(\sum_{i=1}^{m_2} \frac{1}{p_{m_1+i}} \right)^2 \right) \right) \\ &\quad - \binom{m_2+1}{2} \\ &:= F_0(k, m_1, m_2). \end{aligned} \tag{2.11}$$

Let us see how to use the above two counting functions. Let $\ell' < \ell$. Consider ℓ'-tuples

$$\{a_{i_1}, a_{i_2}, \ldots, a_{i_{\ell'}}\} \text{ with } i_1 \le i_2 \le \cdots \le i_{\ell'}.$$

Suppose each $a_{i_t} \in T(a)(k,m)$. Then there are

$$\binom{f(k,m) + \ell' - 1}{\ell'}$$

such ℓ'-tuples. Since a_i's are ℓth power free, this number can be at most ℓ^m. Suppose each $a_{i_t} \in T(a)(k, m_1, m_2)$. Then there are

$$\binom{F(k, m_1, m_2) + \ell' - 1}{\ell'}$$

such tuples. Here each tuple is composed of the first m_1 primes and at most ℓ' prime factors from $p_{m_1+1}, \ldots, p_{m_1+m_2}$. The number of such tuples is at most

$$\ell^{m_1}\binom{m_2 + \ell'}{\ell'}.$$

From this analysis, we get the following lemma (see [1, p. 298–299]).

Lemma 2.2.1 *Suppose* (2.1) *holds wit h* $\ell \geq 3$ *and* $1 \leq \ell' < \ell$.
(i) Suppose $f(k, m) \geq 1$. *Then*

$$\binom{f(k, m) + \ell' - 1}{\ell'} \leq \ell^m.$$

(ii) Suppose $F(k, m_1, m_2) \geq 1$. *Then*

$$\binom{F(k, m_1, m_2) + \ell' - 1}{\ell'} \leq \ell^{m_1}\binom{m_2 + \ell'}{\ell'}.$$

2.3 Largeness of a_i's—A Graph Theoretic Argument

Suppose all the a_i's in $T(a)$ are distinct. Let us arrange the a_i's in increasing order, say

$$a_1' < a_2' < \cdots < a_{|T(a)|}'.$$

Then surely we have $a_i' \geq i$. We want to improve this. Let us consider first $\ell = 2$. Then a_i' are square free. We observe that out of 36 consecutive integers, there are at most 24 square free integers. Let s_h denote the hth square free integer and write it as $s_h = 36f_1 + f_2, 0 \leq f_2 < 36$. So

$$h \leq 24f_1 + \min(f_2, 24) \leq \frac{2}{3}(s_h - f_2) + \min(f_2, 24).$$

Thus $s_h \geq \frac{3}{2}(h - 8)$. So we get

$$a_i' \geq \frac{3}{2}(i - 8) \text{ for } a_i' \in T(a) \text{ if } i \geq 9 \text{ and } |T(a)| \geq 9. \tag{2.12}$$

Let $\ell \geq 3$. Here Erdős and Selfridge [1, p. 296–297] used graph theoretic ideas which have been refined in [2–4]. We explain these ideas now.

Multiplicative Covering. Let $X \geq 1$ and $S_0 \subseteq [1, X]$ be a set of integers. Let U and V be sets of non-negative integers such that every integer in S_0 can be expressed as uv with $u \in U$ and $v \in V$. We call such a pair of sets (U, V), a *multiplicative covering* for S_0. This construction was first given in [1] and it was refined in [2, p. 156–157]. We describe this construction of U and V below. Arrange the distinct elements of S_0 in the increasing order as

$$z_1 < z_2 < \cdots < z_{|S_0|}.$$

Let $S_0(X)$ be the set of z_i's not exceeding X. For positive integers m and T, let $U = U(m, T)$ be the set of integers $\leq T$ which are composed of $p_1, \ldots, p_m$. In other words,

$$U = U(m, T) = \{u : u \geq 1, P(u) \leq p_m\}.$$

Observe that $1 \in U$. Further we understand that an empty product always equals 1.

We construct V as follows. For every prime p_i, $1 \leq i \leq m$, associate an integer $r_i(T)$ such that $p_i r_i(T)$ is the smallest integer $> T$ with $P(p_i r_i(T)) = p_i$. Put

$$r_{m+1}(T) = 1/p_{m+1},\ V_i = \left\{\ p_i w | w \leq X/r_i(T),\ p(p_i w) = p_i\ \right\},\ 1 \leq i \leq m,$$

$$V_{m+1} = \left\{ w | w \leq X,\ p(w) = 1 \text{ or } p(w) \geqq p_{m+1} \right\} \text{ and } V = \cup_{i=1}^{m+1} V_i$$

where $p(n)$ for $n > 1$ denotes the smallest prime divisor of n and we take $p(1) = 1$. The fact that such a pair (U, V) is a multiplicative covering for $S_0(X)$ can be easily checked. We see that for $1 \leq i \leq m+1$,

$$\begin{aligned} |V_i| &= \left| \left\{ w \middle| w \leq \frac{X}{p_i r_i(T)},\ \gcd(w, p_1 \cdots p_{i-1}) = 1 \right\} \right| \\ &= \frac{\varphi(p_1 \cdots p_{i-1})}{p_1 \cdots p_{i-1}} \left\lfloor \frac{X}{p_i r_i(T)} \right\rfloor + E_i, \end{aligned}$$

where E_i's are error terms and φ is the Euler totient function. Since $V_1, \ldots, V_{m+1}$ are pairwise disjoint, we have

$$|V| = \sum_{i=1}^{m+1} \left(\frac{\varphi(p_1 \cdots p_{i-1})}{p_1 \cdots p_{i-1}} \left\lfloor \frac{X}{p_i r_i(T)} \right\rfloor + E_i \right).$$

In [2], it was shown how to calculate the cardinality of V with very good error. Further many sets U were constructed and the corresponding cardinality of V were obtained, see [2, Table 1]. We illustrate this construction with an example here. Let U be the set of all integers ≤ 16 and composed of 2 and 3. Thus $m = 2$, $T = 16$. So $U = \{1, 2, 3, 4, 6, 8, 9, 12, 16\}$. Next $r_1(t) = 16$, $r_2(t) = 6$ and $r_3(t) = 1/5$. Further

$$V_1 = \{2w | 2w \leq X/16\};\ V_2 = \{3w | 3w \leq X/6,\ p(3w) = 3\};$$

$$V_3 = \{w | w \le X, p(w) = 1 \text{ or } p(w) \ge 5\};\ V = V_1 \cup V_2 \cup V_3.$$

In other words,

$$V_1 = \{w | w \le X/32\};\ \ V_2 = \{w | w \le X/18, \gcd(w, 2) = 1\};$$

$$V_3 = \{w | w \le X, \gcd(w, 6) = 1\}.$$

Since V_1 is the set of all integers w with $1 \le w \le X/32$, we get

$$|V_1| \le X/32, \text{ so that } E_1 = 0.$$

Next V_2 is the set of all odd integers $w \le X/18$. Suppose $X = 18X' + f, 0 \le f < 18$, then

$$|V_2| = \begin{cases} X'/2 \text{ if } X' \text{ is even} \\ (X' + 1)/2 \text{ if } X' \text{ is odd.} \end{cases}$$

Thus

$$|V_2| \le \frac{X' + 1}{2} = \left(\frac{X - f}{36}\right) + \frac{1}{2} \le \frac{X}{36} + \frac{1}{2},$$

so that $|E_2| \le 1/2$. Lastly, by inclusion-exclusion principle,

$$|V_3| = X - \lfloor X/2 \rfloor - \lfloor X/3 \rfloor + \lfloor X/6 \rfloor.$$

Let $X = 6X' + f, 0 \le f \le 5$. It is easy to see that

$$|V_3| = \left(1 - \frac{1}{2}\right)\left(1 - \frac{1}{3}\right) X + |E_3|$$

where

$$E_3 = \begin{cases} 0 \text{ if } f = 0, 3 \\ 2/3 \text{ if } f = 1, \\ 1/3 \text{ if } f = 2, 5 \\ -1/3 \text{ if } f = 4. \end{cases}$$

Thus

$$|V_3| \le \frac{X}{3} + \frac{2}{3}.$$

Putting together we have

$$|V| \le \left(\frac{1}{32} + \frac{1}{36} + \frac{1}{3}\right) X + \frac{1}{2} + \frac{2}{3} \le 0.3924X + 2.$$

Bipartite graph. With a given multiplicative cover (U, V) we draw a bipartite graph $G_{S_0(X)} = G_{S_0(X)}(U, V)$ as follows. The vertices of the bipartite graph are the integers in U and the integers in V. Draw an edge between a vertex $u \in U$ and a vertex $v \in V$ if uv equals an integer $r \in S_0(X)$. Thus the number of $z_i \in S_0$ which are $\leq X$ is bounded by the number of edges $E_{S_0(X)}$ of the graph $G_{S_0(X)}$. We proceed to get an upper bound for $E_{S_0(X)}$.

Let ℓ be fixed. Let S_0 consist of ℓth power free integers. We say that the set S_0 satisfies property $P_{\ell'}$ if for any $1 \leq \ell'' \leq \ell' < \ell$, and for any two distinct ℓ''-tuples $(i_1, \ldots, i_{\ell''})$ and $(j_1, \ldots, j_{\ell''})$ with $i_1 \leq \cdots \leq i_{\ell''}$ and $j_1 \leq \cdots \leq j_{\ell''}$, the ratio of the two products $z_{i_1} \cdots z_{i_{\ell''}}$ and $z_{j_1} \cdots z_{j_{\ell''}}$ is not an ℓth power of a rational number. Suppose $S_0(X)$ has all elements distinct. Then $S_0(X)$ has property P_1. Suppose $S_0(X)$ satisfies property $P_{\ell'}$. If $\ell' = 2$, then the graph $G_{S_0(X)}$ will not contain any rectangle. If $\ell' = 3$, then there will not be any rectangle or hexagon. In general, the graph will not contain any $2n$-gon for $n \leq \ell'$.

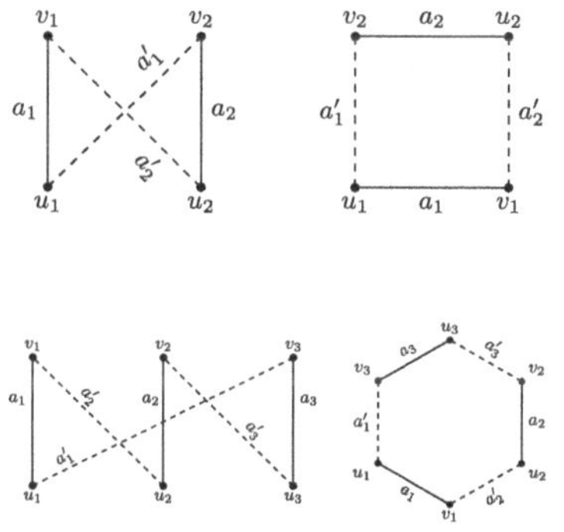

It was shown in [1, Lemma 3] that for $S_0(X)$ with property P_2, the graph $G_{S_0(X)}$ will have the number of edges $E_{S_0(X)}$ bounded by

$$E_{S_0(X)} \leq |V| + \binom{|U|}{2}.$$

In [3, Lemma 4.2], this was improved as

$$E_{S_0(X)} \leq |V| + |W(U)|$$

where $W(U)$ is the set of ratios >1 of pairs of integers from U. This was further improved in [4]. We describe this latest improvement below. For any pair (a, b) with $\gcd(a, b) = 1$ and having prime decomposition $a = r_1^{b_1} \cdots r_s^{b_s}$, $b = s_1^{c_1} \cdots s_t^{c_t}$ where $r_1, \ldots, r_s, s_1, \ldots, s_t$ are primes and b_i, c_j's are positive integers, define the *index* of the pair (a, b) denoted by $Ind(a, b)$ as

$$Ind(a, b) = 1 + b_1 + \cdots + b_s + c_1 + \cdots + c_t.$$

We put $Ind(1, 1) = 1$. The following lemma and its proof are taken from [4, Lemma 2.5].

Lemma 2.3.1 *Let $S_0(X)$ satisfy the property $P_{\ell'}$. Then*

$$E_{S_0(X)} \le |V| + |U^{(0)}|$$

where $U^{(0)}$ is the set of pairs (a, b) of integers from U with $a > b$, $\gcd(a, b) = 1$ and $Ind(a, b)$ either equals 1 or 2 or $> \ell'$.

Proof If a pair of edges emanate from the same vertex $v \in V$, then there exist $u_1, u_2 \in U$ such that $u_1 v$ and $u_2 v$ are elements of S_0. We call such a pair (u_1, u_2) a *concurrent* pair. For $i \ge 1$, let s_i denote the number of vertices in V from which i edges emanate. Then

$$E_{S_0(X)} = \sum_{i\ge1} i s_i = \sum_{i\ge1} s_i + \sum_{i\ge2}(i-1)s_i \le |V| + \sum_{i\ge2}\binom{i}{2}s_i.$$

Let us consider a vertex $v \in V$ from which i edges emanate. The number of concurrent pairs is $\binom{i}{2}$. Thus the total number of concurrent pairs in the graph is

$$\sum_{i\ge2}\binom{i}{2}s_i.$$

In order to estimate this we proceed as follows. Let U consist of integers $\le T$ and composed of $p_1, \ldots, p_m$ where $T > p_m$. Form the set U' of all pairs (a, b) with $a, b \in U, a > b$ and $\gcd(a, b) = 1$. Let

$$U'' = \{(p_1, 1), \ldots, (p_m, 1)\}.$$

Then $U'' \subseteq U'$. Let us assume that all the pairs in U'' are concurrent pairs. Note that these pairs have index 2. Suppose $(a, b) \in U' \backslash U''$ with $2 \le Ind(a, b) \le \ell'$ and it is a concurrent pair. Then there exist integers $v_1, \ldots, v_m, v$ in V such that

$$v_1 = r_{11},\ p_1 v_1 = r_{12}\ ; \ldots ;\ v_m = r_{m1},\ p_m v_m = r_{m2}\ ;\ av = r_a, bv = r_b$$

where $r_{11}, r_{12}, \ldots, r_{m1}, r_{m2}, r_a, r_b$ are elements of $S_0(X)$. Let $a = p_{i_1}^{b_1} \cdots p_{i_s}^{b_s}, b = p_{j_1}^{c_1} \cdots p_{j_t}^{c_t}$. Then we have the following equality.

$$\left(r_{i_1 2}^{b_1} \cdots r_{i_s 2}^{b_s}\right) r_b \left(r_{j_1 1}^{c_1} \cdots r_{j_t 1}^{c_t}\right) = (p_{i_1} v_{i_1})^{b_1} \cdots (p_{i_s} v_{i_s})^{b_s} (bv) \left(v_{j_1}^{c_1} \cdots v_{j_t}^{c_t}\right) =$$

$$\left(v_{i_1}^{b_1} \cdots v_{i_s}^{b_s}\right)\left(p_{i_1}^{b_1} \cdots p_{i_s}^{b_s} v\right)(p_{j_1} v_{j_1})^{c_1} \cdots (p_{j_t} v_{j_t})^{c_t} = \left(r_{i_1 1}^{b_1} \cdots r_{i_s 1}^{b_s}\right)(r_a)\left(r_{j_1 2}^{c_1} \cdots r_{j_t 2}^{c_t}\right).$$

Observe that the left and right extreme elements are product of $1 + \sum_{i=1}^{s} b_i + \sum_{j=1}^{t} c_j \le \ell'$ elements of S_0. This is not possible since $S_0(X)$ has property $P_{\ell'}$. Thus there is no pair $(a, b) \in U' \backslash U''$ with $2 \le Ind(a, b) \le \ell'$ which is concurrent. Let U''' be the set of all pairs (a, b) with $a > b$ and $Ind(a, b) > \ell'$. Let $U^{(0)} = \{(1, 1)\} \cup U'' \cup U'''$. Then the number of concurrent pairs is bounded by $|U^{(0)}|$. This proves the lemma. □

Note. Suppose $\ell' = 2$. Then $U^{(0)}$ is the set of pairs (a, b) of integers from U with $a > b$ and $\gcd(a, b) = 1$. Let $U^c = [1, T] \backslash U$. Then

$$|U^{(0)}| = \sum_{n \in U} \big(\varphi(n) - q(n)\big) \tag{2.13}$$

where $q(n)$ is the number of integers in U^c which are $< n$ and coprime to n.

We illustrate with an example how the above lemma can be used to get a lower bound for a_i's. Let us take $S_0 = T(a)$. Let us assume that $\ell \ge 7$ and S_0 has property P_5, i.e., $\ell' = 5$. Let U be the set of all integers ≤ 16 composed of the primes 2 and 3. As already seen, $|V| \le 0.3924X + 2$. Now, $U^{(0)}$ consists of (a, b) with $a, b \in U$, $a > b$, $\gcd(a, b) = 1$ and $Ind(a, b) = 1, 2$ or > 5 since $\ell' = 5$. So $U^{(0)} = \{(1, 1), (2, 1), (3, 1), (9, 8), (16, 3), (16, 9)\}$. Thus by Lemma 2.3.1,

$$|E_{T(a)(X)}| \le 0.3924X + 8.$$

(U, V) is a multiplicative cover for $T(a)$ and $|E_{T(a)(X)}|$ gives an upper bound for the number of $a_i \le X$. So taking $X = a_h$, we see that

$$h \le 0.3924 a_h + 8$$

which implies that $a_h \ge 2.5484(h - 8)$. Thus it is necessary to know when $P_{\ell'}$ holds.

2.4 Distinctness of a_i's and A_i's

We give below a criterion for the elements of $T(A)$ and $T(a)$ to be distinct.

Lemma 2.4.1 *Assume that* (2.1) *holds with* $a = n$ *and* $|S| = k$. *Suppose* $n > k^{\ell}$ *and* $d \le \ell k^{\ell - 2}$. *Then* $|T(A)| = |T(a)| = k$.

Proof We show that all a_i's are distinct. Suppose not. Then there exist i, j with $0 \le i < j \le k - 1$ such that

$$a_i = a_j.$$

Hence $x_j > x_i$ and

$$\begin{aligned} kd > (j-i)d &= (n+jd)-(n+id) \\ &= a_j x_j^\ell - a_i x_i^\ell = a_i(x_j^\ell - x_i^\ell) \\ &\geq a_i(x_j - x_i)\frac{x_j^\ell - x_i^\ell}{x_j - x_i} \\ &\geq a_i\left(x_j^{\ell-1} + x_j^{\ell-2}x_i + \cdots + x_i^{\ell-1}\right) \\ &\geq \ell a_i x_i^{\ell-1} \geq \ell(a_i x_i^\ell)^{1-\frac{1}{\ell}} \\ &\geq \ell n^{1-\frac{1}{\ell}} > \ell k^{\ell-1} \end{aligned}$$

a contradiction. Similarly all A_i's are distinct. □

Let us write

$$d = d_1 d_2 \text{ with } \gcd(d_1, d_2) = 1 \text{ or } 2.$$

Number of ways this can be done is exactly $2^{\omega(d)+1}$. We shall use this in the next lemma to determine a bound for the number of equal A_i's.

Lemma 2.4.2 *Assume* (2.1) *with* $a = n$, $|S| = k$ *and* $\ell = 2$. *Let* $0 \leq i_0 < k$. *Suppose* $\nu(A_{i_0})$ *denotes the number of distinct* $i \neq i_0$ *satisfying* $A_i = A_{i_0}$. *Further assume that*

$$n + id > k^2 \ for\ every\ i\ with\ A_i = A_{i_0}.$$

Then

$$\nu(A_{i_0}) \leq 2^{\omega(d)+1}.$$

Similar result holds for a_i*'s.*

Proof We prove the result for A_i's. Suppose $A_i = A_{i_0}$ for some $i \neq i_0$. Then

$$|i - i_0|d = |(n+id) - (n+i_0 d)| = A_{i_0}|X_i - X_{i_0}|(X_i + X_{i_0}). \tag{2.14}$$

Note that $\gcd\big((X_i - X_{i_0}), (X_i + X_{i_0}), d\big) = 1$ or 2 since $\gcd(n, d) = 1$. Now $\gcd(d, A_{i_0}) = 1$ implies that $d\big|(X_i - X_{i_0})(X_i + X_{i_0})$. Suppose

$$\nu(A_{i_0}) > 2^{\omega(d)+1} + 1 := Z.$$

This means that there exist at least Z indices i for which $A_i = A_{i_0}$. Then by Box principle, there exists a factorization $d = d_1 d_2$ and indices i_1, i_2 among these Z indices such that

$$d_1\big||X_{i_1} - X_{i_0}|, d_2\big|(X_{i_1} + X_{i_0})\ ;\ d_1\big||X_{i_2} - X_{i_0}|, d_2\big|(X_{i_2} + X_{i_0}).$$

This shows that

$$\frac{d}{\gcd(d_1, d_2)} \Big| (X_{i_1} - X_{i_2}).$$

In particular,

$$|X_{i_1} - X_{i_2}| \geq d/2.$$

Note that $A_{i_1} = A_{i_2}$. So as in (2.14), with (i_0, i) replaced by (i_1, i_2) we get

$$|i_1 - i_2|d = A_{i_1}|X_{i_1} - X_{i_2}|(X_{i_1} + X_{i_2}).$$

With no loss of generality we may take $i_2 > i_1$. Since $A_{i_1} = A_{i_0}$, by assumption we have $n + i_1 d > k^2$. Thus

$$2k > A_{i_1}(X_{i_1} + X_{i_2}) \geq 2A_{i_1}X_{i_1} \geq 2(n + i_1 d)^{1/2} > 2k,$$

a contradiction.

The proof for a_i's is left to the reader. □

When (2.1) holds with $|S| = k, d = 1$ and $\ell > 2$ then one can show that all the products $a_i a_j$ are distinct. In fact, Erdős and Selfridge [1, Lemma 1] proved a stronger result as below which will be used in the proof of Erdős-Selfridge theorem in Chap. 5.

Lemma 2.4.3 *Let* (2.1) *hold with* $a = n, |S| = k, d = 1$ *and* $\ell > 2$. *Suppose the product* $n(n + 1) \cdots (n + k - 1)$ *is divisible by a prime* $> k$. *Then for any* $\ell' < \ell$, *the products* $a_{i_1} \cdots a_{i_{\ell'}} (i_1 \leq \cdots \leq i_{\ell'})$ *are all distinct. In fact, suppose* $(i_1, \ldots, i_{\ell'})$ *and* $(j_1, \ldots, j_{\ell'})$ *with* $0 \leq i_1 \leq i_2 \leq \cdots \leq i_{\ell'} < k$ *and* $0 \leq j_1 \leq j_2 \leq \cdots \leq j_{\ell'} < k$ *are two distinct* ℓ'*-tuples. Then the ratio of the two products* $a_{i_1} \cdots a_{i_{\ell'}}$ *and* $a_{j_1} \cdots a_{j_{\ell'}}$ *is not an* ℓ*th power of a rational number.*

We shall state and prove a result which extends and improves the above lemma considerably. See [5, Lemma 5]. Let us introduce some notation. Write

$$d = D_1 D_2 \tag{2.15}$$

where D_1 is the maximal divisor of d such that all the prime divisors of D_1 are congruent to $1 \pmod{\ell}$. Thus D_1 and D_2 are relatively prime positive integers such that D_2 has no prime divisor which is congruent to $1 \pmod{\ell}$. When $d = 1$ we take $D_1 = D_2 = 1$. Put

$$\theta = \begin{cases} 1 \text{ if } \ell \nmid d \\ 1/\ell \text{ if } \ell | d. \end{cases} \tag{2.16}$$

Lemma 2.4.4 *Let* $1 \leq \ell' \leq \ell - 1, \kappa > 0$ *and*

$$\kappa_0 = \begin{cases} 1 \; if \; d = 1 \\ \min\left(\frac{\ell}{\ell'(\kappa+1)^{(\ell-\ell')/\ell}}, \frac{\kappa}{(\kappa+1)\ell'^{1/\ell}}\right) \; if \; d \neq 1. \end{cases}$$

Assume that (2.1) *holds with* $a = n$. *Suppose* $(n + t_1 d) \cdots (n + t_{|S|} d)$ *is divisible by a prime* $> k$. *Let*

$$D_1 \leq \frac{\kappa_0 \theta}{k}(n + (k-1)d)^{(\ell-\ell')/\ell}.$$

Then for no distinct ℓ'*-tuples* $(i_1, \ldots, i_{\ell'})$ *and* $(j_1, \ldots, j_{\ell'})$ *with* $i_1 \leq \cdots \leq i_{\ell'}$ *and* $j_1 \leq \cdots \leq j_{\ell'}$, *the ratio of the two products* $a_{i_1} \cdots a_{i_{\ell'}}$ *and* $a_{j_1} \cdots a_{j_{\ell'}}$ *is an* ℓ*th power of a rational number.*

Note that when $d = 1$, Lemma 2.4.4 coincides with Lemma 2.4.3.

Proof By the hypothesis,

$$n + t_{|S|} d > k^{\ell}.$$

Suppose there exist two distinct tuples $(i_1, \ldots, i_{\ell'})$ and $(j_1, \ldots, j_{\ell'})$ such that

$$a_{i_1} \cdots a_{i_{\ell'}} = a_{j_1} \cdots a_{j_{\ell'}} (t_1/t_2)^{\ell}$$

where t_1 and t_2 are positive integers. We may assume with no loss of generality that $\gcd(t_1, t_2) = 1$. Put

$$A = \frac{a_{i_1} \cdots a_{i_{\ell'}}}{t_1^{\ell}} = \frac{a_{j_1} \cdots a_{j_{\ell'}}}{t_2^{\ell}}$$

so that A is a positive integer and $\gcd(A, d) = 1$. First we claim that

$$(n + t_{i_1} d) \cdots (n + t_{i_{\ell'}} d) \neq (n + t_{j_1} d) \cdots (n + t_{j_{\ell'}} d).$$

Suppose not. Then $\ell' \geq 2$ and

$$(n + t_{i_1} d) \cdots (n + t_{i_{\ell'}} d) = (n + t_{j_1} d) \cdots (n + t_{j_{\ell'}} d).$$

Cancel any term on the left-hand side which equals some term on the right-hand side. Since $(i_1, \ldots, i_{\ell'}) \neq (j_1, \ldots, j_{\ell'})$, there exists at least one term on the left-hand side different from the initial term $n + t_{j_1} d$ on the right-hand side, say, $n + t_{i_r} d$. Suppose $t_{i_r} > t_{j_1}$. Since $\gcd(n, d) = 1$, we have

$$n + d \leq n + t_{i_r} d \leq \gcd(n + t_{i_r} d, n + t_{j_1} d) \cdots \gcd(n + t_{i_r} d, n + t_{j_{\ell'}} d) \leq k^{\ell'}. \tag{2.17}$$

This means

$$n + t_{|S|} d \leq n + (k-1)d < k(n + d) \leq k^{\ell},$$

a contradiction. This proves our claim. Suppose $t_{i_r} < t_{j_1}$. Then (2.17) holds with $i_r, j_1, \ldots j_{\ell'}$ replaced by $j_1, i_1, \ldots, i_{\ell'}$. So the claim follows. We may now assume that

$$(n + t_{i_1}d) \cdots (n + t_{i_{\ell'}}d) > (n + t_{j_1}d) \cdots (n + t_{j_{\ell'}}d).$$

Then

$$\begin{aligned}(n + t_{i_1}d) \cdots (n + t_{i_{\ell'}}d) &- (n + t_{j_1}d) \cdots (n + t_{j_{\ell'}}d) \\ &= a_{i_1} \cdots a_{i_{\ell'}} x_{i_1}^{\ell} \cdots x_{i_{\ell'}}^{\ell} - a_{j_1} \cdots a_{j_{\ell'}} x_{j_1}^{\ell} \cdots x_{j_{\ell'}}^{\ell} \\ &= \frac{a_{j_1} \cdots a_{j_{\ell'}}}{t_2^{\ell}} \left((t_1 x_{i_1} \cdots x_{i_{\ell'}})^{\ell} - (t_2 x_{j_1} \cdots x_{j_{\ell'}})^{\ell} \right) \\ &= A(x^{\ell} - y^{\ell})\end{aligned} \tag{2.18}$$

where $x = t_1 x_{i_1} \cdots x_{i_{\ell'}}$ and $y = t_2 x_{j_1} \cdots x_{j_{\ell'}}$. Now the left-hand side of (2.18) is divisible by d. Since $\gcd(A, d) = 1, d$ divides $x^{\ell} - y^{\ell}$. It is well known that the prime divisors of

$$\frac{x^{\ell} - y^{\ell}}{x - y}$$

are $\equiv 1 \pmod{\ell}$ or ℓ which divides it to the first power. Thus we see that $x - y$ is divisible by θD_2. Thus $x \geq y + \theta D_2$. (If $d = 1$, then $\theta D_2 = 1$.) Hence the left-hand side of (2.18) is

$$\begin{aligned}&\geq A \left(\binom{\ell}{1} \theta D_2 y^{\ell-1} + \binom{\ell}{2} (\theta D_2)^2 y^{\ell-2} + \cdots + (\theta D_2)^{\ell} \right) \\ &\geq \left(\binom{\ell}{1} \theta D_2 (A y^{\ell})^{(\ell-1)/\ell} + \binom{\ell}{2} (\theta D_2)^2 (A y^{\ell})^{(\ell-2)/\ell} + \cdots + (\theta D_2)^{\ell} \right) \\ &\geq \left(\binom{\ell}{1} \theta D_2 n^{\ell'(\ell-1)/\ell} + \binom{\ell}{2} (\theta D_2)^2 n^{\ell'(\ell-2)/\ell} + \cdots + (\theta D_2)^{\ell} \right),\end{aligned}$$

since $A y^{\ell} = (a_{j_1} x_{j_1}^{\ell}) \cdots (a_{j_{\ell'}} x_{j_{\ell'}}^{\ell}) \geq n^{\ell'}$. On the other hand, the left-hand side of (2.18) is

$$\begin{aligned}&< (n + kd)^{\ell'} - n^{\ell'} \\ &= \binom{\ell'}{1} kd n^{\ell'-1} + \binom{\ell'}{2} (kd)^2 n^{\ell'-2} + \cdots + (kd)^{\ell'}.\end{aligned}$$

Comparing the above two estimates, we obtain

$$\begin{aligned}&\left\{ \binom{\ell}{1} \theta D_2 n^{\ell'(\ell-1)/\ell} - \binom{\ell'}{1} kd n^{\ell'-1} \right\} \\ &+ \left\{ \binom{\ell}{2} (\theta D_2)^2 n^{\ell'(\ell-2)/\ell} - \binom{\ell'}{2} (kd)^2 n^{\ell'-2} \right\} \\ &+ \cdots + \left\{ \binom{\ell}{\ell'} (\theta D_2)^{\ell'} n^{\ell'(\ell-\ell')/\ell} - (kd)^{\ell'} \right\} + \cdots + (\theta D_2)^{\ell} < 0.\end{aligned} \tag{2.19}$$

Let $d = 1$. Then $\theta D_2 = 1$ and $n > k^\ell$. Hence every term in the curly brackets is >0. This is a contradiction. Next let $d > 1$. We consider two cases.

Case 1. Let $n \geq \frac{n+(k-1)d}{\kappa+1}$. For $1 \leq i \leq \ell'$, the ith term in the curly bracket in (2.19) is

$$\begin{aligned}
&\binom{\ell}{i}(\theta D_2)^i n^{\ell'(\ell-i)/\ell} - \binom{\ell'}{i}(kd)^i n^{\ell'-i} \\
&= n^{\ell'-i} D_2^i \left\{\binom{\ell}{i}\theta^i n^{i(\ell-\ell')/\ell} - \binom{\ell'}{i} k^i D_1^i\right\} \\
&\geq 0
\end{aligned}$$

since by assumption,

$$D_1 \leq \frac{\ell}{\ell'(\kappa+1)^{(\ell-\ell')/\ell}} \frac{\theta}{k}(n+(k-1)d)^{(\ell-\ell')/\ell}$$

and

$$\binom{\ell}{i} \geq \binom{\ell'}{i}\left(\frac{\ell}{\ell'}\right)^i.$$

This contradicts (2.19).
Case 2. Let $n < \frac{n+(k-1)d}{\kappa+1}$. Then

$$\begin{aligned}
d &\geq \frac{1}{k}(n+(k-1)d-n) \\
&\geq \frac{1}{k}\left(n+(k-1)d - \frac{n+(k-1)d}{\kappa+1}\right) \\
&= \frac{\kappa}{k(\kappa+1)}(n+(k-1)d).
\end{aligned}$$

From (2.19), it follows that there exists an i with $1 \leq i \leq \ell'$ such that

$$\binom{\ell}{i}(\theta D_2)^i n^{\ell'(\ell-i)/\ell} - \binom{\ell'}{i}(kd)^i n^{\ell'-i} < -\frac{(\theta D_2)^\ell}{\ell'}.$$

This gives

$$\binom{\ell'}{i}(kd)^i n^{\ell'-i} > \frac{(\theta D_2)^\ell}{\ell'},$$

which implies

$$D_1^\ell > \frac{\theta^\ell d^{\ell-i}}{\ell'\binom{\ell'}{i}k^i n^{\ell'-i}}.$$

Using the upper bound for n and the lower bound for d, we get

$$\begin{aligned}
D_1^\ell &> \frac{\theta^\ell d^{\ell-i}}{\ell'\binom{\ell'}{i}k^i n^{\ell'-i}} \\
&> \left(\frac{\theta\kappa}{k}\right)^\ell \frac{1}{\ell'(\kappa+1)^{\ell-\ell'}} \frac{1}{\kappa^i\binom{\ell'}{i}}(n+(k-1)d)^{\ell-\ell'} \\
&> \left(\frac{\theta\kappa}{k}\right)^\ell \frac{1}{\ell'(\kappa+1)^\ell}(n+(k-1)d)^{\ell-\ell'}.
\end{aligned}$$

Thus

$$D_1 > \frac{\theta}{k}\frac{\kappa}{\ell'^{1/\ell}(\kappa+1)}(n+(k-1)d)^{(\ell-\ell')/\ell}$$

which contradicts our assumption on D_1. This completes the proof of the lemma. □

Remark. Under the hypothesis of Lemma 2.4.4, $T(a)$ has property $P_{\ell'}$. From the condition on D_1, it is clear that we need good lower bounds for $n+(k-1)d$. By the hypothesis we know that

$$n+(k-1)d \geq p_{\pi(k)+1}^\ell \geq (k+1)^\ell.$$

Can we improve this further? This will be an important step in many problems.

Exercise 2

(1) Justify (2.11).
(2) Prove that the number of ℓ'-tuples $(a_{i_1}, \ldots, a_{i_{\ell'}})$ is

$$\binom{f(k,m)+\ell'-1}{\ell'}$$

if $a_{i_t} \in T(a)(k,m)$, $1 \leq t \leq \ell'$. Get an upper bound for the number of such tuples if $a_{i_t} \in T(a)(k,m_1,m_2)$.
(3) Show that the sets U and V constructed in Sect. 2.3 is a multiplicative cover for $S_0(X)$.
(4) Write a computer program to calculate the errors E_i in Sect. 2.3 and find the errors E_1 to E_{11}. It was shown in [2, Lemma 7] that

$$E_1 \leq 0,\ E_2 \leq \frac{1}{2},\ E_3 \leq \frac{2}{3},\ E_4 \leq \frac{14}{15},\ E_5 \leq \frac{53}{35},$$

$$E_6 \leq \frac{194}{77},\ E_7 \leq \frac{3551}{1001},\ E_8 \leq \frac{92552}{17017}$$

$$E_9 \leq \frac{2799708}{323323},\ E_{10} \leq \frac{9747144}{676039},\ E_{11} \leq \frac{58571113}{2800733}.$$

Can you improve these values?

(5) Prove (2.13).

(6) In how many ways can one write an integer $d > 1$ as $d = d_1 d_2$ with $\gcd(d_1, d_2) = 1$ or 2.

(7) Let $x > y$ be integers and assume $\ell > 2$ is a prime. Show that the prime divisors of

$$\frac{x^\ell - y^\ell}{x - y}$$

are $\equiv 1 \pmod{\ell}$ or ℓ which divides it to the first power.

(8) Modify Lemma 2.4.1 to show that A_i's are distinct and so are a_i's if

$$D_1 \leq (0.84)\theta k^{\ell-2}$$

whenever $n + (k-1)d > k^\ell$. (We will use this in Chap. 6.)

References

1. P. Erdős, J.L. Selfridge, The product of consecutive integers is never a power. Illinois J. Math. **19**, 292—301 (1975). [**2, 5**]
2. N. Saradha, On perfect powers in products with terms from arithmetic progressions. Acta Arith. **82**, 147–172 (1997). [**2–5**]
3. N. Saradha, R. Tijdeman, Arithmetic progressions with common difference divisible by small primes. Acta Arith. **131**, 267–279 (2008). [**2**]
4. N. Saradha, Rational points with large denominator on Erdős-Selfridge superelliptic curves. Publ. Math. Debrecen **99**, 317–329 (2021). [**2**]
5. N. Saradha, T.N. Shorey, Almost perfect powers in arithmetic progression. Acta Arith. **138**, 363–388 (2001). [**2, 6, 7**]

Chapter 3
Theorem of Sylvester-Extensions and Refinements

$$P(2 \times 9 \times 16) = 3;\ P(18 \times 25 \times 32) = 5;\ P(64 \times 81 \times 98) = 7$$

The well-known Bertrand's postulate states that there is a prime between x and $2x$ for any real number $x > 1$. A proof of this fact was later found by Chebyshev. Since then finding prime in *short* intervals has been the center problem in Prime Number Theory. This is a difficult problem. We may relax this problem and ask if a sequence of integers can have an integer which has a large prime divisor. To make the problem more tractable, one may restrict to consecutive integers or integers in an arithmetic progression. A positive assertion and proof was first given by Sylvester [1] in 1892. In his own words, his proof is *wearisome.* Schur [2] rediscovered and proved Sylvester's result again in 1929. In the case of consecutive integers the proofs were simplified by Erdős [3] in 1934. Sylvester's theorem for arithmetic progression was given a complete stature by Shorey and Tijdeman [4] in 1990. In this chapter, we shall state Sylvester's theorem for consecutive integers and give the proof of Erdős (Theorem 3.1.1). For the case of arithmetic progression, we shall present the result of Shorey and Tijdeman (Theorem 3.1.2). The proof given here is self-contained in the sense that it does not assume previously known results for arithmetic progressions of small length. We apply Sylvester's theorem to prove the irreducibility of truncated exponential polynomials (Theorem 3.4.1). Some extensions and refinements of Sylvester's theorem are given in Sect. 3.5.

3.1 Statement of the Results

Notation. Let n, d and k denote positive integers. Further let

$$\Delta(n, d, k) = n(n + d) \cdots (n + (k - 1)d)$$

S. Natarajan, *Perfect Powers—An Ode to Erdős*, Infosys Science Foundation Series,
https://doi.org/10.1007/978-981-96-2599-4_3

and put

$$\Delta(n, 1, k) = \Delta(n, k).$$

In this notation, our problem is to find out how big $P(\Delta(n, d, k))$ can be. When $k = 1$,

$$\Delta(n, d, 1) = n$$

and anything can happen. To avoid such trivialities, we always assume that $k \geq 2$.

Theorem 3.1.1 *Suppose $n > k$. Then*

$$P(\Delta(n, k)) > k.$$

Remarks.

(1) The condition $n > k$ is necessary since we always have $P(\Delta(1, k)) \leq k$.
(2) When $n = k + 1$, we have

$$\Delta(n, k) = (k + 1) \cdots (2k).$$

The theorem says that there is an integer between $k + 1$ and $2k$, say m with $P(m) > k$. When $k = 2$, we may take $m = k + 1 = 3$. For $k \geq 3$, m must be a prime since $m \leq 2k$. Thus m is a prime between $k + 1$ and $2k$ and this is Bertrand's postulate.
(3) Suppose $P(\Delta(n, k)) \leq k$. Then by Theorem 3.1.1 we have $n \leq k$. Further none of $k + 1, \ldots, k + n - 1$ can be a prime. Hence $\pi(n + k - 1) = \pi(k)$. Converse is obvious. Thus $P(\Delta(n, k) \leq k$ if and only if $\pi(n + k - 1) = \pi(k)$. Thus this problem is equivalent to the well-known problem of gaps between primes. This is a rich area of activity in Prime Number Theory. See also Comment 1 below.

Next we take the case $d > 1$, i.e., product of successive terms from an arithmetic progression. We show

Theorem 3.1.2 *Let $d > 1, k > 2$ and $\gcd(n, d) = 1$. Then*

$$P(\Delta(n, d, k)) > k$$

except when $(n, d, k) = (2, 7, 3)$.

Remarks.

(1) Let $k = 2$. Then $\Delta(n, d, 2) = n(n + d)$. Suppose $P(n(n + d)) \leq 2$. This means both n and $n + d$ are pure powers of 2. Since $\gcd(n, d) = 1$ we have $n = 1$ and $d = 2^h - 1$ for some integer $h \geq 2$. We avoid this case.
(2) Sylvester proved that $P(\Delta(n, d, k)) > k$ if $n \geq d + k$. This was improved to $n > k$ by Langevin [5]. In Theorem 3.1.2 which is due to Shorey and Tijdeman this restriction is completely removed and the only exception is $(n, d, k) = (2, 7, 3)$.

(3) The proofs depend on many computations and these are carried out using MATHEMATICA. The reader may use any other mathematical package available nowadays like SAGE.

3.2 Proof of Theorem 3.1.1

We assume that $n > k$. Put $x = n + k - 1$. Then $x \geq 2k$. Further

$$\frac{\Delta(n,k)}{k!} = \frac{x(x-1)\cdots(x-k+1)}{k!} = \binom{x}{k}.$$

Thus Theorem 3.1.1 is equivalent to showing that

$$P\left(\binom{x}{k}\right) > k \text{ if } x \geq 2k. \tag{3.1}$$

We now suppose the contrary that

$$x \geq 2k, \text{ but } P\left(\binom{x}{k}\right) \leq k. \tag{3.2}$$

Step 1. We first show that under the above supposition, x can be bounded above in terms of k. To be precise,

$$x < \begin{cases} k^2 \text{ for } k \leq 32, k \neq 3, 5, 7 \\ k^{3/2} \text{ for } k \geq 33. \end{cases} \tag{3.3}$$

It is clear that

$$\binom{x}{k} = \frac{x(x-1)\cdots(x-k+1)}{k(k-1)\cdots 1} > \left(\frac{x}{k}\right)^k$$

and

$$\binom{x}{k} = \prod_{\substack{p \leq k \\ p\text{-prime}}} p^{\nu_p\left(\binom{x}{k}\right)} \leq x^{\pi(k)}$$

by Lemma 1.5.1. Thus

$$x < k^{\frac{k}{k-\pi(k)}}.$$

Now we use Theorem 1.1.1 for an upper bound for $\pi(k)$. Then we get

$$\frac{k}{k-\pi(k)} \le \frac{1}{1-\frac{1}{\log k}-\frac{1.2762}{(\log k)^2}}.$$

Note that the right-hand side expression is a monotonic decreasing function, say $f(k)$ of k. Hence if $f(k) < 3/2$ for some $k = k'$, then $f(k) < 3/2$ for all $k \ge k'$. We find the least such k'. Call it k_0. In this case, we find $k_0 = 53$. So $\frac{k}{k-\pi(k)} \le 3/2$ for $k \ge 53$. The value of k_0 can still be brought down by using the exact value of $\pi(k)$ for $k \le 52$. For instance, when $k = 52$, $\pi(k) = 15$ so that

$$\frac{k}{k-\pi(k)} < 1.41 < 3/2.$$

We conclude the same for $33 \le k \le 52$. Further for $k \le 32$, $\frac{k}{k-\pi(k)} \le 2$ except for $k = 3, 5, 7$. Thus

$$\frac{k}{k-\pi(k)} \le \begin{cases} 2 \text{ if } k \le 32, k \ne 3, 5, 7 \\ 3/2 \text{ if } k \ge 33. \end{cases}$$

This proves (3.3). The reader will meet with this kind of computation at several places in the book. At many places we may follow this process without any mention.

Step 2. Let $k \ge 33$. By **Step** 1, we may assume that $x < k^{3/2}$. We reduce this bound further.

(A) We show that $x < 4k$. We shall use the fact that

$$\binom{2k}{k} \ge \frac{4^k}{2k}.$$

Suppose that $x \ge 4k$. Then by Corollary 1.5.4, we get

$$\binom{4k}{k} \le \binom{x}{k} < 4^{k+\sqrt{x}} < 4^{k+k^{3/4}}. \tag{3.4}$$

On the other hand,

$$\binom{4k}{k} = \binom{2k}{k}\frac{4k(4k-1)\cdots(3k+1)}{2k(2k-1)\cdots(k+1)} > \frac{4^k \cdot 2^k}{2k} = \frac{8^k}{2k}.$$

Thus by (3.4), we have

$$\frac{8^k}{2k} < 4^{k+k^{3/4}}$$

and so

$$\frac{2^k}{2k} < 4^{k^{3/4}}$$

or

$$2^{k-2k^{3/4}-1} < k$$

which implies that $k \le 34$. For $k = 33, 34$ we check directly that (3.4) does not hold. This contradiction proves that $x < 4k$ whenever $k \ge 33$.

(B) Suppose $\lceil 5k/2 \rceil < x < 4k$. Then we shall show that $k \le 106$. As in (A), using Corollary 1.5.4, we have

$$\binom{\lceil 5k/2 \rceil}{k} \le \binom{x}{k} < 4^{k+\sqrt{x}} < 4^{k+2\sqrt{k}} \tag{3.5}$$

and

$$\binom{x}{k} \ge \binom{2k}{k} \frac{\frac{5}{2}k(\frac{5}{2}k-1)\cdots(\frac{5}{2}k-k+1)}{2k(2k-1)\cdots(k+1)} > \frac{4^k}{2k}\left(\frac{5}{4}\right)^k.$$

So

$$\left(\frac{5}{4}\right)^k < 2k4^{2\sqrt{k}} < 2^{2\sqrt{k}}4^{2\sqrt{k}} = 2^{6\sqrt{k}},$$

as $2k < 2^{2\sqrt{k}}$. Since $(5/4)^4 > 2$, we infer that

$$2^{k/4} < 2^{6\sqrt{k}}$$

which gives $k \le 575$. To reduce this even further, use (3.5) and get $k \le 106$.

(C) Let $2k \le x \le \lfloor 5k/2 \rfloor$. Then we shall show that $k \le 114$. We have

$$\frac{4^k}{2k} \le \binom{2k}{k} \le \binom{x}{k} \tag{3.6}$$

In this case we show that $P\left(\binom{x}{k}\right) \le x/3$. By Assumption (3.2), all prime factors p of $\binom{x}{k}$ are $\le k$. Suppose $x/3 < p \le x$. Then

$$2k/3 < p \le k, x - k \ge 2k - k = k \ge p$$

and

$$x - k \le x - 2x/5 = 3x/5 < 9p/5 < 2p.$$

So

$$\nu_p(x!) = 2, \nu_p(k!) = 1 \text{ and } \nu_p((x-k)!) = 1.$$

So none of the primes p with $x/3 < p \le x$ divide $\binom{x}{k}$. Hence

$$P\left(\binom{x}{k}\right) \le x/3.$$

Then

$$\binom{x}{k} \le 4^{x/3+\sqrt{x}}.$$

So from (3.6), it follows that

$$\frac{4^k}{2k} \le \binom{2k}{k} < 4^{x/3+\sqrt{x}} < 4^{5k/6+\sqrt{5k/2}} \tag{3.7}$$

giving

$$4^{k/6} < 2k4^{1.59\sqrt{k}} < 4^{2.59\sqrt{k}}$$

or

$$4^{.166k-2.59\sqrt{k}} < 1$$

which implies that $k \le 243$. Now use (3.7) itself to get $k \le 114$.

In conclusion, by **Step 1** and **Step 2(A)–(C)**, we find that

$$\begin{cases} k \le 32 \text{ and } 2k \le x < k^2 \\ 33 \le k \le 114 \text{ and } 2k \le x < 4k. \end{cases} \tag{3.8}$$

We are left with the finitely many values of x and k in (3.8). For these values we directly check that (3.2) is not valid, in other words, Eq. (3.1) is valid for these values. To do this we use the following commands in MATHEMATICA.

(i) To find $P(\binom{x}{k})$-define

$$G[x, k] := \text{First[Last[FactorInteger[Binomial}[x, k]]]].$$

Then $G[x, k] = P(\binom{x}{k})$.

(ii) To check if $P(\binom{x}{k}) > k$ for $k_1 \le k \le k_2$ and $x_1(k) \le x \le x_2(k)$-use

$$\text{For}[k = k_1, k \le k_2, k++,\ \ \text{For}[x = x_1(k), x \le x_2(k), x++,$$

$$\text{If}[G[x, k] \le k, \text{Print}[(x, k)]]]]$$

Apply (i) and (ii) with $k_1 = 2, k_2 = 32, x_1(k) = 2k$ and $x_2(k) = k^2$ first and then with $k_1 = 33, k_2 = 114, x_1(k) = 2k$ and $x_2(k) = 4k$ to find that no value of (x, k) is printed. Thus $P(\binom{x}{k}) > k$ for these values of k and x. Thus (3.1) is valid for all the values in (3.8). This completes the proof of Theorem 3.1.1. □

Remark. In this era of computations, we assume that the reader is familiar with some of the mathematical packages. So we will not be giving the commands as above at other places.

3.3 Proof of Theorem 3.1.2

This is the first time we will be applying *the fundamental argument of Erdős* and clearly the reader can understand the strength of the argument. Let us assume that $k \geq 3$ and

$$P(\Delta(n, d, k)) \leq k. \tag{3.9}$$

We shall arrive at a contradiction. We apply the fundamental argument of Erdős from Chap. 2. In that notation,

$$a = n, S = \{n, n + d, \ldots, n + (k - 1)d\}, n_i = n + id = u_i$$

by our Assumption (3.9). So by (2.4), there exists a subset $S_1 \subset S$ with $|S_1| \geq k - \pi_d(k)$ such that

$$\prod_{i=0}^{k-\pi_d(k)-1} (n + id) \leq \prod_{u_i \in S_1} u_i \leq (k - 1)! \prod_{\substack{p \mid d \\ p \leq k-1}} p^{-\nu_p(k-1)!}. \tag{3.10}$$

On the other hand,

$$\prod_{i=0}^{k-\pi_d(k)-1} (n + id) \geq (k - \pi_d(k) - 1)! d^{k-\pi_d(k)-1}$$

and

$$\prod_{i=0}^{k-\pi_d(k)-1} (n + id) \geq (k - \pi_d(k))! d^{k-\pi_d(k)} \text{ if } n \geq d.$$

Hence from (3.10),

$$\begin{aligned} d^{k-\pi_d(k)-1} &\le (k-1)(k-2)\cdots(k-\pi_d(k)) \prod_{\substack{p|d \\ p\le k-1}} p^{-\nu_p(k-1)!} \\ &\le (k-1)(k-2)\cdots(k-\pi_d(k)) \prod_{\substack{p|d \\ p\le k-1}} p^{-\left(\frac{k-1}{p}+\frac{k-1}{p^2}+\frac{k-1}{p^3}-3\right)} \\ &< k^{\pi_d(k)} \prod_{\substack{p|d \\ p\le k-1}} p^{-\left(\frac{k-1}{p}+\frac{k-1}{p^2}+\frac{k-1}{p^3}-3\right)} \\ &< k^{\pi_d(k)}. \end{aligned} \tag{3.11}$$

and if $n \ge d$,

$$\begin{aligned} d^{k-\pi_d(k)} &\le (k-1)(k-2)\cdots(k-\pi_d(k)+1) \prod_{\substack{p|d \\ p\le k-1}} p^{-\nu_p(k-1)!} \\ &\le (k-1)(k-2)\cdots(k-\pi_d(k)+1) \prod_{\substack{p|d \\ p\le k-1}} p^{-\left(\frac{k-1}{p}+\frac{k-1}{p^2}+\frac{k-1}{p^3}-3\right)} \\ &< k^{\pi_d(k)-1} \prod_{\substack{p|d \\ p\le k-1}} p^{-\left(\frac{k-1}{p}+\frac{k-1}{p^2}+\frac{k-1}{p^3}-3\right)} \\ &< k^{\pi_d(k)-1}. \end{aligned} \tag{3.12}$$

First Case. We show that under (3.9), $d \ge 6$. Suppose $k \ge 3$ and $d \in \{2, 3, 4, 5\}$. First let $d = 2, 4$. Then $\pi_d(k) \le \pi(k) - 1$. Use (3.11) to get

$$\begin{aligned} d &\le (k^{\pi_d(k)} 2^{-(1/2+1/4+1/8)(k-1)} 2^3)^{1/(k-\pi_d(k)-1)} \\ &\le (k^{(\pi(k)-1)} 2^{-(7/8)(k-1)} 2^3)^{1/(k-\pi(k))}. \end{aligned}$$

Now we use upper estimate for $\pi(k)$ from Theorem 1.1.1 to get

$$d \le \exp\left(\frac{1 + \frac{1.2762}{\log k} - \frac{\log k}{k} - \frac{7\log 2}{8}(1 - 1/k) + \frac{3\log 2}{k}}{1 - \frac{1}{\log k} - \frac{1.2762}{\log^2 k}} \right).$$

This is not valid for $k \ge 995$ if $d = 2$ and for $k \ge 24$ if $d = 4$. Thus we conclude that

$$k \le 994 \text{ if } d = 2;\ k \le 23 \text{ if } d = 4.$$

Now we use (3.11) with exact value of $\pi(k)$ to conclude that $d \neq 4$ and

$$d = 2, k = 19, 23, 24, 31, 32, 47.$$

To exclude these values of k assume $n \geq 2 (= d)$ and apply (3.12). This enables us to conclude that $n = 1$. Then it is easy to check that (3.10) is not valid. Thus $d \neq 2$.

Let $d = 3$. The procedure is as above with $p = 3$. For $k \geq 4$, $\pi_d(k) \leq \pi(k) - 1$. Using (3.11) with Theorem 1.1.1, first we conclude that $k \leq 71$. With exact value of $\pi(k)$ we exclude all these values of k. Let $k = 3$. By (3.9), $\gcd(n, d) = 1$ and $d = 3$, we get

$$P(n(n+3)(n+6)) \leq 2$$

and so (3.10) implies that

$$n(n+3) \leq 2$$

which is not valid for $n \geq 1$. Thus $d \neq 3$.

Let $d = 5$. By the procedure above with $p = 5$ we conclude that $k \in \{3, 4, 5\}$. Further using (3.12), we get $n \leq 4$. It is a simple check to see that in these few cases there is always a prime $> k$ dividing $\Delta(n, 5, k)$. Thus $d \neq 5$.

Second case. Let $k \geq 9$. We use the following fact.

$$\pi_d(k) \leq \pi(k) \leq (k-1)/2 \text{ for } k \geq 9.$$

So by (3.11), we get $d \leq k - 1$. Hence $\pi_d(k) \leq \pi(k) - 1$. Further

$$d \leq k^{\pi_d(k)/(k-\pi_d(k)-1)} \leq k^{(\pi(k)-1)/(k-\pi(k))}.$$

Now we use upper estimate for $\pi(k)$ from Theorem 1.1.1 to get

$$d \leq \exp\left(\frac{1 + \frac{1.2762}{\log k} - \frac{\log k}{k}}{1 - \frac{1}{\log k} - \frac{1.2762}{\log^2 k}}\right).$$

Note that the right-hand side is a monotonic decreasing function of k. We find that

$$d \leq 5 \text{ for } k \geq 69.$$

Further (3.11) implies

$$d^{k-\pi(k)} \leq (k-1)(k-2)\cdots(k-\pi(k)+1).$$

We use this with exact values of $\pi(k)$ for $9 \le k \le 68$ to find that

$$d \le 5 \text{ for } k \ge 9$$

which does not hold by First case. Thus $k \le 8$.

Third case. Let $3 \le k \le 8$ and $d \ge 6$. By (3.10), $n(n+d) \le 6, 24$ for $k = 4, 5$. This is possible only when

$$k = 5 : n = 1, 6 \le d \le 23;\ n = 2, 6 \le d \le 10. \tag{3.13}$$

Similarly, we derive that

$$\begin{cases} k = 6 : n = 1, d = 6, 7; \\ k = 7 : n = 1, 6 \le d \le 18;\ n = 2, 6 \le d \le 11, \\ \qquad n = 3, 6 \le d \le 8;\ n = 4, d = 6; \\ k = 8 : n = 1, d = 6, 7, 8;\ n = 2, d = 6. \end{cases} \tag{3.14}$$

For the values of n, d, k given in (3.13) and (3.14), it is checked directly that (3.9) is not valid as in the proof of Theorem 3.1.1.(Note that here we have one more variable d.) This shows that $k = 3$. Then by (3.10), $n \le 2$. Let $n = 1$. Then by (3.9),

$$P((d+1)(2d+1)) \le 3.$$

So

$$d + 1 = 2^{\alpha_1} 3^{\beta_1};\ 2d + 1 = 2^{\alpha_2} 3^{\beta_2}$$

which gives

$$2^{\alpha_1 + 1} 3^{\beta_1} - 2^{\alpha_2} 3^{\beta_2} = 1.$$

This implies that $\alpha_2 = 0$ and $\beta_1 = 0$. Then

$$2^{\alpha_1 + 1} - 3^{\beta_2} = 1.$$

If $\alpha_1 \ge 2$, reducing $\pmod 8$ we get $3^{\beta_2} \equiv 7 \pmod 8$ which is not possible. Thus $\alpha_1 = 1$ which gives $d = 1$, not possible. Let $n = 2$. Then d is odd and

$$P(2(2+d)(2+2d)) = 3$$

or

$$P((d+1)(d+2)) = 3.$$

Since $\gcd((d+1),(d+2)) = 1$ we have

$$d + 1 = 2^\alpha, d + 2 = 3^\beta, \alpha \geq 2$$

so that

$$3^\beta = 2^\alpha + 1$$

giving β even, by taking (mod 4). Then

$$2^\alpha = (3^{\beta/2} - 1)(3^{\beta/2} + 1).$$

Since $\gcd((3^{\beta/2} - 1), (3^{\beta/2} + 1)) = 2$, we get

$$3^{\beta/2} - 1 = 2, 3^{\beta/2} + 1 = 2^{\alpha-1}.$$

Then $3^{\beta/2} = 3$ giving $\beta = 2$, so that $d = 7$ and we have the one exception $(n, d, k) = (2, 7, 3)$! □

Remark. The whole proof depended on the one inequality (3.10) which is a consequence of the fundamental argument of Erdős and also on many simple computations.

3.4 An Application of Sylvester's Theorem

In this section we shall apply Theorem 3.1.1 to show the irreducibility of exponential polynomials. In [2], Schur rediscovered Sylvester's theorem and applied it to determine the irreducibility of many orthogonal polynomials. Here we give a proof in the spirit of Schur's. We refer to Sury [6, p. 29–31] for this elegant proof of Schur.

Theorem 3.4.1 *For any positive integer n the truncated exponential polynomial*

$$E_n(X) = n!\left(1 + X + \frac{X^2}{2!} + \cdots + \frac{X^n}{n!}\right)$$

is irreducible over $\mathbb{Z}[X]$.

Proof We will be using the following fact from algebraic number theory. Suppose K is a number field. Any non-zero ideal I in the ring of integers $\mathcal{O}_K$ of K can be expressed uniquely up to order as a product of prime ideals $\mathfrak{p}$. Further if $\nu_\mathfrak{p}(I)$ denotes the power with which $\mathfrak{p}$ appears in the prime factorization of I, then

$$\nu_\mathfrak{p}(I) \leq [K : \mathbb{Q}]$$

where $[K : \mathbb{Q}]$ denotes the degree of K over $\mathbb{Q}$. Let us write

$$E_n(X) = c_0 + c_1 X + \cdots + X^n$$

with $c_i = n!/i!$. If possible, let

$$E_n(X) = f(X)g(X) \text{ with } f(X), g(X) \in \mathbb{Z}[X].$$

Since $E_n(X)$ is monic, f and g are also monic. Let

$$f(X) = a_0 + a_1 X + \cdots + X^r \text{ and } g(X) = b_0 + b_1 X + \cdots + X^{n-r}.$$

We can also assume without loss of generality that $f(X)$ is irreducible and $r \leq n/2$.

Step 1. Let p be a prime dividing a_0, the constant term in $f(X)$. We show that $p \leq \deg f = r$. Let α be a root of $f(X)$ and let $K = \mathbb{Q}(\alpha)$. Note that the norm of α, denoted as $\mathcal{N}(\alpha)$ satisfies

$$\mathcal{N}(\alpha) = \pm a_0.$$

Let (α) and (p) be the ideals generated by α and p, respectively, in K. Since $p|a_0$, there is a prime ideal $\mathfrak{p}$ of $\mathcal{O}_K$ such that

$$(\alpha) = \mathfrak{p}^k I, (p) = \mathfrak{p}^\ell J, k \geq 1, \ell \geq 1$$

and I, J are ideals with $\gcd(I, J) = 1$. Further $\ell \leq \deg K = r$. Since α is a root of $E_n(X)$ we have

$$\alpha^n + n\alpha^{n-1} + \cdots + \frac{n!}{2!}\alpha^2 + n!\alpha = -n! \tag{3.15}$$

We know

$$\nu_p(n!) = \left\lfloor \frac{n}{p} \right\rfloor + \left\lfloor \frac{n}{p^2} \right\rfloor + \cdots := h_n, say.$$

Note that $h_n \leq n/(p-1)$. Hence the power with which $\mathfrak{p}$ divides the ideal $(n!)$ in $\mathcal{O}_K$ is given by

$$\nu_{\mathfrak{p}}(n!) = \ell h_n.$$

Similarly, the power of $\mathfrak{p}$ in the ideals formed by each of the terms on the left-hand side of (3.15) is

$$\nu_{\mathfrak{p}}\left(\frac{n!\alpha^i}{i!}\right) = \ell(h_n - h_i) + ki > 0 \text{ for } 1 \leq i \leq n.$$

By (3.15), *not* every term can have power of p strictly exceeding ℓh_n. So there exists some i with $1 \leq i \leq n$ such that

$$\ell(h_n - h_i) + ki \leq \ell h_n$$

or

$$ki \leq \ell h_i.$$

Thus

$$i \leq ki < \ell i/(p-1)$$

giving $p - 1 < \ell \leq r$ and so $p \leq r$ as required.

Step 2. We show that there exists a prime $q > r$ dividing $c_0, c_1 \ldots, c_{n-r}$. Here is where we will be using Sylvester's theorem. Note that

$$c_{n-r} = (n-r+1)\cdots n$$

is a product of r consecutive terms and the initial term $n - r + 1 > r$ since $r \leq n/2$. Hence by Theorem 3.1.1, there exists a prime $q > r$ dividing c_{n-r} and obviously therefore q divides $c_0, c_1 \ldots, c_{n-r-1}$.

Step 3. We reduce $E_n(X) = f(X)g(X) \pmod q$ and get a contradiction. This should remind the reader of Eisenstein's criterion for irreducibility of polynomials. By Step 2,

$$E_n(X) = c_{n-r+1}X^{n-r+1} + \cdots + X^n \pmod q.$$

Now

$$f(X)g(X) = a_0b_0 + (a_1b_0 + a_0b_1)X + \cdots + (a_{n-r}b_0 + \cdots + a_0b_{n-r})X^{n-r} + \cdots$$

By the expression $E_n(X) \pmod q$, it follows that

$$a_0b_0 \equiv 0 \pmod q, a_1b_0 + a_0b_1 \equiv 0 \pmod q, \ldots,$$

$$a_{n-r}b_0 + \cdots + a_0b_{n-r} \equiv 0 \pmod q.$$

Since $a_0 \not\equiv 0 \pmod q$ (why?), we have $b_0 \equiv 0 \pmod q$ and recursively we get

$$b_0 \equiv b_1 \equiv \cdots \equiv b_{n-r} \equiv 0 \pmod q.$$

This is impossible since $b_{n-r} = 1$. □

3.5 Some Extensions and Refinements of Sylvester's Theorem

Since the simplified proof of Sylvester's theorem by Erdős, several variations of the theorem have been considered by Erdős himself and others. We shall mention below a few. There are mainly three aspects to Sylvester's theorem, viz.,

(1) The initial term $= n > k$,
(2) $P(\Delta(n, d, k)) > k$,
and
(3) $k =$ Length of $\Delta(n, d, k)$.

Let us analyze these aspects.

(1) Can we improve the condition $n > k$? This is possible provided k is large. We give some motivation below. As noted in Remark 3 after Theorem 3.1.1, the problem of finding for which integers n and k we have $P(\Delta(n, k)) \leq k$ is equivalent to finding when

$$\pi(n + k - 1) = \pi(k),$$

i.e., there is no prime between k and $n + k - 1$. There are several results in the literature on this question. A classical, but yet unsolved question about primes is Legendre's conjecture, viz.,

Conjecture (Legendre): *There exists a prime p between any two consecutive squares, i.e., for all $x \geq 1$ there exists a prime p such that*

$$p \in \left(x^2, (x + 1)^2\right).$$

The length of the interval here is $2x + 1$ which is *small* compared to the initial term x^2. Generally, we can find the following Bertrand type results.

There exists a prime $p \in (x, (1 + \delta)x)$ for $x \geq x_0(\delta)$
which can be derived from Prime Number Theorem with error term,

$$\pi(x) = \frac{x}{\log x} + O\left(\frac{x}{\log^2 x}\right).$$

Taking $x = k$ and $\delta x = n$, such results would imply that if $n > \delta k$, then $P(\Delta(n, k)) > k$. For instance, it follows by re-phrasing a result of Kadiri-Lumley [7] that

$$P(\Delta(n, k)) > k \text{ if } n > k/2442159713 \text{ for } k \geq 1.4 \times 10^{65}.$$

Consider now short intervals of the form $(x, x + x^{\theta})$ for $\theta \in (0, 1)$. The best result so far with smallest θ is due to Baker et al. [8].

Theorem 3.5.1 (Baker-Harman-Pintz) *There is a prime in the interval* $(x, x + x^{0.525})$ *for sufficiently large* x.

The value $\theta = 0.525$ is significant because the size of the interval $(x, x + x^{0.525})$ is close to the size of the interval

$$(x, x + 2x^{0.5} + 1)$$

predicted by Legendre's conjecture. Applying Theorem 3.5.1 to our problem, we get that

$$P(\Delta(n, k)) > k$$

even if $n > k^{0.525}$ when k is large. It is difficult to determine how large k should be because of the inherent nature of the proofs in Analytic Number Theory.

(2) Can we improve $P(\Delta(n, d, k)) > k$? It is expected that as we increase the lower bound k there may be some exceptions and the set of exceptions will increase as the lower bound increases. Let $d = 1$. One of the early results in this direction is due to Hanson [9] who showed that

$$P(\Delta(n, k)) > 1.5k \text{ for } (n, k) \notin \{(3, 2), (8, 2), (6, 5)\}.$$

There have been several improvements of such a result by Shorey, Tijdeman and their co-authors. In [10, Corollary 1], Laishram and Shorey showed that

$$P(\Delta(n, k)) > 1.8k \text{ for } n > k > 2$$

except for an explicitly given set of values of (n, k). Such results involve meticulous application of estimates involving $\pi(x)$ and computation to make the exceptional set explicit. Let $d > 1$. As in Theorem 3.1.2, it can be shown that

$$\omega(\Delta(n, d, k)) \geq \pi(k), \tag{3.16}$$

see [11]. A conjecture of Schinzel, known as Hypothesis H, implies that

there are infinitely many d for which both $1 + d$ *and* $1 + 2d$ *are primes.*

Thus the above inequality is best possible for $k = 3$. Now $\Delta(n, d, k)$ has $\pi(k) - \pi_d(k)$ primes $\leq k$ dividing it. By the above result, therefore, there are atleast $\pi_d(k)$ primes $> k$ dividing $\Delta(n, d, k)$. Hence

$$P(\Delta(n, d, k)) \geq p_{\pi(k)+\pi_d(k)}.$$

It follows from estimates for nth prime p_n that the right-hand side is much bigger than k. For instance, when $k = 10$, $d = 3$, we have $\pi_d(k) = 3$ and $p_{\pi(k)+\pi_d(k)} = p_7 = 17 > k$. Thus improving (3.16) has consequences on Sylvester's theorem. Moree [12] sharpened (3.16) as

$$\omega(\Delta(n,d,k)) > \pi(k) \text{ if } k \geq 4 \text{ and } (n,d,k) \neq (1,2,5).$$

He conjectured that

$$\omega(\Delta(n,d,k)) \geq \pi(2k) - 1 \text{ if } k \geq 4,\ (n,d,k) \neq (1,3,10).$$

This was proved by Laishram and Shorey [13, Corollary 1]. We refer to an expository article of Shorey [14] for more results.

(3) Let us define $f(k)$ as the least integer so that the product of $f(k)$ consecutive integers each greater than k always contains a prime greater than k. Theorem 3.1.1 gives that

$$f(k) \leq k.$$

Can we shorten the length k? In 1955, Erdős [15, Theorem 1] showed the following result.

Theorem 3.5.2 *There is an absolute constant $c_1 > 1$ so that*

$$f(k) \leq c_1 \frac{k}{\log k}.$$

In other words,

$$P(n(n+1)\cdots(n+t)) > k \text{ if } t = \left\lfloor \frac{c_1 k}{\log k} \right\rfloor \text{ and } n > k.$$

By results of Rankin [16], for every k there are consecutive primes p_r and p_{r+1} satisfying $k < p_r < p_{r+1} < 2k$. So it seems likely that $f(k)$ will not be substantially larger than the greatest difference between p_r and p_{r+1}. Further by [16] and the result of Shorey [17, Theorem 3], what we know so far is that

$$\frac{\log k(\log\log k)(\log\log\log\log k)}{(\log\log\log k)^2} \ll f(k) \ll \frac{k \log\log\log k}{\log k(\log\log k)}.$$

Following a famous conjecture of Cramer, one expects that the order of magnitude of $f(k)$ is $(\log k)^2$. Erdős felt that it is hopeless to prove or disprove this. He suggested that it would be interesting to compute $f(k)$ for $k < 100$. Many mathematicians computed the values of $f(k)$ and at present it is known for $k \leq 178$. See [18, Theorem 1.1]. For instance, if $k = 178$, then $f(k) = 14$. Thus if $n > 178$, then

$$P(n(n+1)\cdots(n+13)) > 178.$$

We end this section by proving Theorem 3.5.2.

Proof of Theorem 3.5.2. We will be assuming at many places that k is large to satisfy the inequalities involved. Let us use Theorem 3.5.1 for

$$n+1, \ldots, n+t, t = \left\lfloor c_1 \frac{k}{\log k} \right\rfloor$$

with $n+1 = x, t = (n+1)^{0.525} = \left\lfloor c_1 \frac{k}{\log k} \right\rfloor$. So if $n \le k^{3/2}$ (note that $3/2 < 1/(0.525)$), at least one of the integers $n+1, \ldots, n+t$ is a prime, for sufficiently large c_1. So we need to consider the case $n > k^{3/2}$. Further by Theorem 3.1.1, we may assume that $t < k$. Suppose Theorem 3.5.2 is false. Then for any $c_1 > 0$ we can find a $n > k^{3/2}$ so that

$$P((n+1)\cdots(n+t)) \le k.$$

Hence

$$P\left(\binom{n+t}{t}\right) \le k.$$

Now we apply Lemma 1.5.1 to get

$$\binom{n+t}{t} \le (n+t)^{\pi(k)}.$$

On the other hand,

$$\binom{n+t}{t} = \frac{(n+1)\cdots(n+t)}{t!} \ge \left(\frac{n+t}{t}\right)^t > \left(\frac{n}{t}\right)^t.$$

Thus

$$\left(\frac{n}{t}\right)^t \le (n+t)^{\pi(k)}.$$

Now we use

$$t^t < k^t < n^{2t/3}, n+t < n+k < n+n^{2/3} < n^2$$

and

$$\pi(k) \le \frac{3k}{2\log k}$$

(note that we are using only a weaker estimate for $\pi(k)$) to obtain

$$n^{t/3} < n^{3k/\log k}.$$

This is a contradiction for $c_1 > 9$. This proves Theorem 3.5.2. □

The above theorem also suggests that there may be several terms in $n+1, \ldots, n+k$ divisible by primes $> k$ when $n > k$ and k large. As seen in Problem (2) above, this will have effect on $P(\Delta(n, k))$. Erdős showed the following result.

Let $n > k^{3/2}$ and k very large. Then there are at least $k/6$ integers in $(n+1), \ldots, (n+k)$ having prime factors $> k$.

The proof of this is similar to the proof of Theorem 3.5.2. As an immediate consequence, we get

$$P(\Delta(n+1, k)) \geq p_{\pi(k)+k/6} \gg k \log k. \tag{3.17}$$

3.6 Comments

(1) In [2, 19], Schur established four theorems concerning the irreducibility of certain classes of orthogonal polynomials over the rationals. We saw in Theorem 3.4.1 the case of exponential polynomials. In fact Schur also computed the Galois group of this polynomial as A_n, the alternating group of n letters if $4|n$ and S_n, the symmetric group of n letters, otherwise. In [20], Coleman gave a proof of Schur's result using *Newton Polygons*. See also [21]. In the early 2000s, Filaseta and his co-authors generalized the theorems of Schur and their works depended on the utility of Newton Polygons. More work was done by Shorey and his co-authors in the later part of the decade.

(2) We saw that improvement on $P(\Delta(n, k))$ depended on estimates of $\pi(x)$, Sylvester's theorem and fundamental argument of Erdős. Further improvement on this came from Ramachandra, Jutila, Shorey and Tijdeman. The present status is as follows.

$$P(\Delta(n, k)) \gg k \log k \frac{\log \log k}{\log \log \log k} \quad \text{for } \log k > e.$$

Their results depended on method of Roth and Halberstam on difference between consecutive k-free integers, estimates for exponential sums and best estimates for linear forms in logarithms of algebraic numbers close to 1. In the case $d > 1$, Shorey and Tijdeman [22, Theorem 2] showed that

$$P(\Delta(n, d, k)) \gg_\epsilon k \log \log \chi \text{ if } \chi > k(\log k)^\epsilon, \epsilon > 0$$

where $\chi = n + (k-1)d$. We refer to the survey article of Shorey [23] and the references given there.

Exercise 3

(1) Show that (3.1) is true for any k if it is true whenever k is a prime.
(2) Using the fundamental argument of Erdős show that (3.1) is true for $k = 2, 3, 5, 7$.
(3) Use Erdős argument to show that $\omega(\Delta(n, d, k)) \geq \pi(k)$.
(4) Use the argument in the First Case of Proof of Theorem 3.1.2 to exclude $d \leq 10$.
(5) Prove (3.17) and the result of Erdős used in it.
(6) Show the irreducibility of the exponential polynomial using the well-known Eisenstein's criterion.
(7) Show that any polynomial can be written as sum of two irreducible polynomials.

References

1. J.J. Sylvester, On arithmetical series. Messenger of Math, bf 21 (1892), 1–19, 87–120; and Collected mathematical papers, **4**, 687–731 (1912). [**3**]
2. J. Schur, Einige Satze über Primzahlen mit Anwendung auf Irreduzibilitätsfragen. Sitzungsberichteder preussischen Akademieder Wissenschaften, Phys. Math. Klasse **23**, 1–24 (1929). [**3**]
3. P. Erdős, A theorem of Sylvester and Schur. J. Lond. Math Soc. **9**, 282–288 (1934). [**1, 3**]
4. T.N. Shorey, R. Tijdeman, On the greatest prime factor of an arithmetical progression, in *A Tribute to Paul Erdős*, ed. by A. Baker, B. Bollobás, A. Hajnal (Cambridge University Press, 1990), pp. 385–389. [**3**]
5. M. Langevin, Plus grand facteur premier d'entiers en progression arithmétiques, Séminaire Délange-Pisot–Poitou, **18** e année, no.3, 7p, (1976/77)
6. B. Sury, *Number Theory and Combinatorics* (Indian Academy of Sciences Publications, 2017), pp. 186. [**3**]
7. H. Kadiri, A. Lumley, Short effective intervals containing primes. Integers **14**, Paper No. A61, 18 (2014). [**3**]
8. R.C. Baker, G. Harman, J. Pintz, The difference between consecutive primes, II. Proc. Lond. Math. Soc. **83**(3), 532–562 (2001). [**3**]
9. D. Hanson, On a theorem of Sylvester and Schur. Canad. Math. Bull. **16**, 195–199 (1973). [**3**]
10. S. Laishram, T.N. Shorey, The greatest prime divisor of a product of consecutive integers. Acta Arith. **120**(3), 299–306 (2005). [**3**]
11. T.N. Shorey, R. Tijdeman, On the number of prime factors of a finite arithmetical progression. Siachun Daxue Xuebao **26**, 72–74 (1989). [**3**]
12. P. Moree, On arithmetical progressions having only few different prime factors in comparison with their length. Acta Arith. **70**, 295–312 (1995). [**3**]
13. S. Laishram, T.N. Shorey, Number of prime divisors in a product of terms of an arithmetic progression. Indag. Math., N. S., **15**(4), 505–521 (2004). [**3**]
14. T.N. Shorey, Mathematical contributions. Bull. Bombay Math. Colloquium **15** (1999) published in 2001, 1–19. [**3**]
15. P. Erdős, On consecutive integers. Nieuw Archief voor Wiskunde (3) III, 124–128 (1955). [**3**]
16. R. Rankin, The difference between consecutive prime numbers. J. Lond. Math. Soc. **13**, 242–247 (1938). [**3**]
17. T.N. Shorey, On gaps between numbers with a large prime factor II. Acta Arith. **25**, 365–373 (1974). [**3**]
18. M. Bauer, M. Bennett, Prime factors of consecutive integers. Math. Comp. **77**, 2455–2459 (2008). [**3**]

19. J. Schur, Einige Satze über Primzahlen mit Anwendung auf Irreduzibilitätsfragen II. Ibid, 370–391. [**3**]
20. R.F. Coleman, On the Galois groups of the exponential Taylor polynomials. L'Enseignement Mathématique, t. **33**, 183–189 (1987). [**3**]
21. T.N. Shorey, *Theorems of Sylvester and Schur*. The Mathematics Student, Special Centenary, vol. 2007 (2007), pp. 135–145. [**3**]
22. T.N. Shorey, R. Tijdeman, On the greatest prime factor of an arithmetical progression (III), in *Diophantine Approximation and Transcendental Numbers*, ed. by P. Philippon, W. de Gruyter, Luminy, 1990 (New York, 1992), pp. 275–280. [**3**]
23. T.N. Shorey, Exponential Diophantine equations involving products of consecutive integers and related equations, in *Number Theory*, ed. by R.P. Bambah, V.C. Dumir, R.J. Hans-Gill, Hindustan Book Agency (1999), pp. 463–495; Trends Math. (Birkhäuser, Basel, 2000), pp. 463–495. [**3, 8**]

Chapter 4
Theorem of Erdős and Rigge and Some Generalizations

$\binom{50}{3} = 140^2$; $\frac{6!}{5} = 12^2$; $\frac{10!}{7} = 720^2$

In 1724, Goldbach wrote to Bernoulli that a product of *three* consecutive positive integers is not a square, thereby starting a series of investigations by mathematicians. In 1939, Rigge [1] and Erdős [2], independently, showed that a product of *two or more* consecutive positive integers is never a square. In other words, the Diophantine equation

$$n(n+1)\cdots(n+k-1) = y^2 \text{ in integers } n > 0, k > 1, y \geq 1$$

is impossible or in our notation $\Delta(n, k)$ is never a square. In fact, Rigge proved much more. Let us consider the equation

$$\Delta(n, k) = by^2 \text{ with integer } b \geq 1.$$

Rigge showed that this equation is not possible if $P(b) \leq k/2$. In 1951, Erdős [3] proved the impossibility of this equation when $b = k!$ and $k \geq 4$. When $k = 3$ there is a solution, viz., $48 \cdot 49 \cdot 50 = 3!(140)^2$. It is suggestive that we can consider the equation with $P(b) \leq k$, i.e.,

$$\Delta(n, k) = by^2 \text{ with } P(b) \leq k. \tag{4.1}$$

We shall show the impossibility of this equation for $k \geq 4$ (Theorem 4.1.1). A variation of the above equation is considered in Sect. 4.3. Next we turn to an analogue of (4.1) for arithmetic progressions. Let us consider

$$\Delta(n, d, k) = by^2, \text{ with } P(b) \leq k \text{ and } d > 1. \tag{4.2}$$

S. Natarajan, *Perfect Powers—An Ode to Erdős*, Infosys Science Foundation Series,
https://doi.org/10.1007/978-981-96-2599-4_4

In 1985, Marszalek [4] showed that

Equation (4.2) *with* $b = 1$ *implies that* $k \le 2\exp(d(d+1)^{1/2})$.

In other words, for a fixed $d > 1$, Eq. (4.2) with $b = 1$ implies that k is bounded by an absolute constant. In 1990, Shorey and Tijdeman [5] improved the above result. In this chapter, we shall describe their result (Theorem 4.1.2).

4.1 Statement of the Results

The first result is taken from Saradha [6, Theorem 1].

Theorem 4.1.1 *Equation (4.1) with* $n > k$ *and* $k \ge 4$ *has no solution.*

We have seen in Chap. 3 that $n > k$ is necessary in Sylvester's theorem. Otherwise, there are infinitely many values of k for which $P(\Delta(n,k)) \le k$ for instance when $n = 1$. Then (4.1) also has infinitely many solutions. Hence we need the condition $n > k$. Then by Sylvester's theorem, there exists a prime $p > k$ dividing a term say, $n + i$ of $\Delta(n,k)$. Now (4.1) implies that $p|y$ and hence $p^2|(n+i)$. So

$$n = (n+i) - i > p^2 - k \ge (k+1)^2 - k > k^2.$$

Our next result is due to Shorey and Tijdeman on (4.2). See [5, Theorem 1(a)].

Theorem 4.1.2 *There exist effectively computable positive absolute constants* C_1 *and* C_2*, such that (4.2) with* $k > C_1$ *implies that*

$$2^{\omega(d)} > C_2 \frac{k}{\log k}.$$

Here $\omega(d)$ *denotes the number of distinct prime divisors of* d.

Remark.

(i) Note that the above theorem is an improvement over the result of Marszalek since in place of d we have $\omega(d)$.
(ii) By (1.4), it follows that

$$d \ge k^{C_3 \log\log k}$$

where $C_3 > 0$ is an effectively computable absolute constant.
(iii) By Theorem 3.1.2, we know that $\Delta(n,d,k)$ is divisible by a prime $> k$ except when $(n,d,k) = (2,7,3)$. Thus we have

$$n + (k-1)d \ge (k+1)^2.$$

An essential ingredient in their proof is an improvement of the above inequality.

4.2 Proof of Theorem 4.1.1

We begin with two lemmas which are left as exercises to the reader. The first lemma was proved by A. Meyl in 1878. Erdős quotes this in [3]. See [7, p. 25]. The second lemma is due to Euler.

Lemma 4.2.1 *The only solutions of the equation*

$$n(n+1)(n+2) = 6y^2$$

in integers $n > 1, y > 1$ *are* $n = 2, y = 2;\ n = 48, y = 140$.

Lemma 4.2.2 *Product of four consecutive positive integers is never a square.*

Proof of Theorem 4.1.1 We use the notation in the fundamental argument of Erdős and the representation of n_i from Chap. 2. Let $a = n, d = 1, t_i = i$ and $|S| = k$ so that by (2.6),

$$n + i = a_i x_i^2,\ P(a_i) \le k,\ a_i \text{ square free},\ 0 \le i \le k-1.$$

Since $n > k$ as seen earlier, $n > k^2$. We will be dealing with the product

$$\prod_{i=0}^{k-1} a_i.$$

By Lemma 2.4.1, all a_i's are distinct since $n > k^2$.

Step 1. We apply a mild variation of fundamental argument of Erdős. We know that $P(a_i) \le k$ for $0 \le i < k$. Since a_i's are square free, there are at most $\lceil \frac{k}{p} \rceil$ such a_i's divisible by p for any prime $p < k$. If $p = k$, it can divide at most one a_i. Thus

$$\prod_{i=0}^{k-1} a_i \,\Big|\, k \prod_{p<k} p^{\lceil \frac{k}{p} \rceil}.$$

Now

$$\left\lceil \frac{k}{p} \right\rceil \le \frac{k-1}{p} + 1 \le \nu_p((k-1)!) + 1.$$

So

$$\prod_{p<k} p^{\lceil \frac{k}{p} \rceil} \le (k-1)! \prod_{p<k} p.$$

Hence

$$\prod_{i=0}^{k-1} a_i \,\Big|\, (k-1)! \prod_{p\le k} p \tag{4.3}$$

Therefore it follows from (1.14) that

$$\prod_{i=0}^{k-1} a_i \leq (k-1)!(2.78)^k.$$

Since a_i's are distinct, we have

$$\prod_{i=0}^{k-1} a_i \geq k!. \tag{4.4}$$

The reader can very well see that this lower bound will not suffice to get a contradiction with the above upper bound. First we reduce the upper bound. We use the idea in (2.5) with $q = 2$ and 3. From (2.8) we have

$$\alpha_2 = \nu_2\left(\prod_{i=0}^{k-1} a_i\right) \leq \frac{k}{3} + \frac{2\log k}{\log 2}; \ \alpha_3 = \nu_3\left(\prod_{i=0}^{k-1} a_i\right) \leq \frac{k}{4} + \frac{2\log k}{\log 3}$$

and

$$\nu_p((k-1)!) \leq \frac{k}{p-1}.$$

Thus from (4.3)

$$\prod_{i=0}^{k-1} a_i \leq \frac{(k-1)!(2.78)^k}{2^{k-\alpha_2}3^{\frac{k}{2}-\alpha_3}}$$

which gives

$$\begin{aligned}\prod_{i=0}^{k-1} a_i &\leq (k-1)!(2.78)^k 2^{-2k/3+2\log k/\log 2} 3^{-k/4+2\log k/\log 3}\\ &\leq (k-1)!(1.34)^k k^4.\end{aligned}$$

Still this is not sufficient to get a contradiction with the lower bound (4.4).

Step 2. Let $k \geq 9$. We improve the lower bound in (4.4). We shall use the fact that a_i's are square free. We have $|T(a)| = k \geq 9$. So from (2.12) we get

$$\prod_{i=0}^{k-1} a_i \geq \prod_{i=0}^{k-1} s_i \geq (1\cdot 2\cdot 3\cdot 5\cdot 6\cdot 7\cdot 10\cdot 11)\left(\frac{3}{2}\right)^{k-8}(k-8)!$$

Step 3. We claim $k \leq 590$. Comparing the upper and lower bounds found in **Steps** 1 and 2, we get

$$138600(1.11)^k \leq k^{11}\left(\frac{3}{2}\right)^8$$

which gives

$$5407(1.11)^k \leq k^{11}.$$

This is not possible for $k \geq 591$. This shows our claim.

Step 4. We use the counting argument in Sect. 2.2 to exclude $9 \leq k \leq 590$. For this, we compute $f_0(k, m)$ given in (2.9). Below we specify $(m, f_0(k, m))$ when $9 \leq k \leq 590$.

$$\begin{cases} (2, 5) \text{ for } 9 \leq k \leq 22;\ (3, 9) \text{ for } 23 \leq k \leq 106; \\ (4, 17) \text{ for } 107 \leq k \leq 286;\ (5, 33) \text{ for } 287 \leq k \leq 556; \\ (6, 65) \text{ for } 557 \leq k \leq 590. \end{cases}$$

This gives the necessary contradiction to (2.10) since $f_0(k, m) > 2^m$ in all the cases. So we conclude that $k \leq 8$.

Step 5. We exclude $4 \leq k \leq 8$. Recall that all a_i' s are distinct. So if there are four a_i's in $\Delta(n, k)$ with $P(a_i) \leq 3$, then they must be from $\{1, 2, 3, 6\}$. Hence the product of these four terms is a square and so by Lemma 4.2.2, they cannot be consecutive terms in $\Delta(n, k)$. Let us call this property as PFT. By this, we first see that $k \neq 4$. It is enough to look for the possibilities when $f(k, 2) \leq 4$ and exclude them by congruence properties.

Let $k = 6$. Then $f_0(k, 2) \geq 5$ except when $5|a_0$ and $5|a_5$. In this case, applying property PFT to $n+1, n+2, n+3, n+4$ we get a contradiction.

Let $k = 8$. Then $f_0(k, 2) \geq 5$ except when

$$7|a_0, 7|a_7 \text{ and } 5|a_1, 5|a_6.$$

Application of property PFT to $n+2, n+3, n+4, n+5$ rules out this possibility.

Let $k = 5$. We may assume that $f_0(k, 2) = 4$. Further by property PFT we may assume that $5|a_1$ or $5|a_2$ or $5|a_3$. We use Legendre symbol $(\frac{\cdot}{\cdot})$. to exclude these possibilities. Let $5|a_1$. Then $n \equiv 4 \pmod 5$. Hence

$$\left(\frac{n}{5}\right) = 1, \left(\frac{n+2}{5}\right) = 1, \left(\frac{n+3}{5}\right) = -1, \left(\frac{n+4}{5}\right) = -1.$$

Using gcd properties we find that

$$n = 6x_0^2, n+2 = x_2^2, n+3 = 3x_3^2, n+4 = 2x_4^2.$$

Apply Lemma 4.2.1 to $(n+2)(n+3)(n+4) = 6y^2$ to exclude this case. Next let $5|a_2$. Then $n \equiv 3 \pmod 5$. Hence

$$\left(\frac{n}{5}\right) = -1, \left(\frac{n+1}{5}\right) = 1, \left(\frac{n+3}{5}\right) = 1, \left(\frac{n+4}{5}\right) = -1.$$

This implies $n+1 = x_1^2, n+3 = 6x_3^2$ or $n+1 = 6x_1^2, n+3 = x_3^2$. Then $n+4 = x_4^2$ or $n = x_0^2$ which contradicts distinctness of a_i's (or one can use $\left(\frac{n}{5}\right) = -1, \left(\frac{n+4}{5}\right) = -1$). Lastly, $5|a_3$. Then $n \equiv 2 \pmod 5$. Then

$$\left(\frac{n}{5}\right) = -1, \left(\frac{n+1}{5}\right) = -1, \left(\frac{n+2}{5}\right) = 1 \left(\frac{n+4}{5}\right) = 1.$$

We argue as in the first case (because of symmetry) and exclude this last case.

Let $k = 7$. Then $f_0(k, 2) \geq 5$ except when there are three distinct integers $0 \leq i_1, i_2, i_3 \leq 6$ such that $7|a_{i_1}, 5|a_{i_2}, 5|a_{i_3}$. By property PFT, we may assume that

$$(i_1, i_2, i_3) \in \{(1,0,5), (2,0,5), (2,1,6), (3,0,5), (3,1,6), (4,0,5),$$
$$(4,1,6), (5,1,6)\}.$$

Let us consider $(i_1, i_2, i_3) = (1, 0, 5)$. Then $n \equiv 6 \pmod 7$. So

$$\left(\frac{n+2}{7}\right) = 1, \left(\frac{n+3}{7}\right) = 1, \left(\frac{n+4}{7}\right) = -1 \left(\frac{n+6}{7}\right) = -1.$$

This implies that $(n+4, n+6) \in \{3x^2, 6y^2\}$ which is not possible. For the remaining cases, we give below the necessary details by which these cases are excluded.

$$(ii) - (2,0,5) : n \equiv 5 \pmod 7, (n+3, n+4, n+6) \in \{x^2, 2y^2\}.$$

$$(iii) - (2,1,6) : n \equiv 5 \pmod 7, (n, n+5) \in \{3x^2, 6y^2\}.$$

$$(iv) - (3,0,5) : n \equiv 4 \pmod 7, (n+1, n+2) \in \{3x^2, 6y^2\}.$$

$$(v) - (3,1,6) : n \equiv 4 \pmod 7, (n, n+4, n+5) \in \{x^2, 2y^2\}.$$

$$(vi) - (4,0,5) : n \equiv 3 \pmod 7, (n+2, n+3) \in \{3x^2, 6y^2\}.$$

$$(vii) - (4,1,6) : n \equiv 3 \pmod 7, (n+2, n+3) \in \{3x^2, 6y^2\}.$$

$$(viii) - (5,1,6) : n \equiv 2 \pmod 7, (n+3, n+4) \in \{3x^2, 6y^2\}.$$

To exclude the cases (*ii*) and (v), distinctness of a_i's is used. The remaining cases are excluded by gcd conditions. This completes the proof of Theorem 4.1.1. □

Remark. The above proof is typical of the methods used in these kinds of problems. As a first step one bounds k from above. For the remaining finitely many values of k one needs to argue differently using congruences or other methods which the reader will come across as we go on.

4.3 A Variation of Eq. (4.1)

In this section, instead of the full product of k consecutive integers, we will consider a subproduct of integers taken from a set of consecutive integers. Let

$$S = \{n + t_1 < n + t_2 < \cdots < n + t_{|S|}\}$$

be a set of $|S| \le k$ integers belonging to $\{n, n+1, \ldots, n+k-1\}$. In the previous section we have taken $|S| = k$. Let us now take $|S| < k$. The question is how small we can take $|S|$ so that equation

$$(n + t_1)(n + t_2) \cdots (n + t_{|S|}) = by^2,\ P(b) \le k \tag{4.5}$$

has no solution. In 1955, Erdős [8] observed that the proof given in Sect. 4.1 is sufficient to show the following result.

Suppose (4.5) *holds with* $n > k^2$. *Then there exist absolute constants* C_4 *and* C_5 *such that*

$$\text{if } |S| \ge k - C_4 \frac{k}{\log k},\ \textit{then } k \le C_5.$$

In other words, under the given hypothesis, Eq. (4.5), which is a product obtained from $\Delta(n, k)$ by omitting $C_4 \frac{k}{\log k}$ terms has no solution for $k > C_5$. In [6, Eqs. (2) and (3)], it was shown that if $C_4 < (0.0156)$ then one may take $C_5 = 870$. A closer look at Steps 1–5 makes it clear that we cannot replace the constant C_4 with a function $f(k) \to \infty$ as $k \to \infty$. Thus we need a different argument if we want to omit more terms than of order $k/\log k$. In 1987, Shorey [9] used Baker's theorem on integral solutions of hyper-elliptic equations and Sieve-theoretic arguments to show the following result.

Theorem 4.3.1 *Let* $\epsilon > 0$. *Suppose (4.5) holds with* $n > k^2$ *and*

$$|S| \ge k - (1 - \epsilon)\frac{k \log \log k}{\log k}.$$

Then k *is bounded by a number depending only on* ϵ.

We will prove this theorem in this section. Erdős proposed this in [8]. There is a marginal improvement of the condition on $|S|$ by Balasubramanian and Shorey (see [10]) as

$$|S| \geq k - k\left(\frac{\log\log k}{\log k} - \frac{\log\log\log k}{\log k} - \frac{\theta}{\log k}\right)$$

where θ is any real number. They also showed that if $|S|$ is taken less than the quantity on the right-hand side then there are products which happen to be squares. An analogous result for (4.2) was proved by Saradha and Shorey [11].

Proof of Theorem 4.3.1. Let all the hypotheses of the theorem hold. Let $\epsilon > 0$ be given. We assume that k is very large depending on ϵ, say $k > k_0(\epsilon)$ so that the inequalities which are encountered below are valid. We use the notation from Sect. 2.2. Put $T(a) = \{a_1, \ldots, a_{|S|}\}$. All a_i's are distinct since $n > k^2$ so that $|T(a)| = |S|$. Further as seen in Step 1 of Theorem 4.1.1,

$$\prod_{i=1}^{|S|} a_i \leq (k-1)!(2.78)^k < (2.78k)^k. \tag{4.6}$$

Step 1. Let $S_1 = \{a_i \in S, a_i \leq k(\log k)^{1-\epsilon/2}\}$. We show that

$$|S_1| \geq \epsilon k/2 \text{ and } x_i > k^{1/4} \text{ for } a_i \in S_1. \tag{4.7}$$

To see this, let t be the number of a_i' s in S with $a_i > k(\log k)^{1-\epsilon/2}$. Then by (4.6),

$$e^{t(\log k + (1-\epsilon/2)\log\log k)} \leq e^{k\log k + 1.03k}$$

which gives $t < (1-\epsilon)k$. Thus

$$|S_1| \geq |S| - (1-\epsilon)k \geq \epsilon k - (1-\epsilon)\frac{k\log\log k}{\log k} \geq \epsilon k/2.$$

For any $a_i \in S_1$ we find that

$$x_i^2 = (n+t_i)/a_i > k(\log k)^{-1+\epsilon/2} > k^{1/2}.$$

This gives $x_i > k^{1/4}$.

Step 2. Let $S_2 = \{a_i \in S_1, a_i \leq 3k\}$. Then we show that

If $|S_2| > \epsilon k/4$, *then* k *is bounded by a number depending only on* ϵ.

For this first we apply sieve argument of Erdős as in Lemma 1.4.2, form a superelliptic equation and then apply Theorem 1.2.3. Let $|S_2| > \epsilon k/4$. We take the set B in Lemma 1.4.2 with $z = k/\delta^2$ and $X = 3k$ where $\delta > e^{24/\epsilon}$. Then

$$|B| \ll 3k/\log(3\delta^2) < \epsilon k/16.$$

Let

$$S_3 = \{a_i : a_i \in S_2 \text{ and } a_i \geq k/\delta^2\}.$$

Then

$$|S_3| \geq |S_2| - k/\delta^2 > \epsilon k/8.$$

As seen in the proof of Lemma 1.4.3, every $a_i \in S_3$ is divisible by at least one b_i. Suppose every b_i appears in at most two elements of S_3 then

$$|S_3| \leq 2|B| < \epsilon k/8$$

a contradiction. So we may assume that there exists a $b_i \in B$ such that it divides three distinct elements from S_3. Without loss of generality, we may assume that these are a_1, a_2 and a_3. We use the following simple identity.

$$(n + t_2)(n + t_3) = (n + t_1 + (t_2 - t_1))(n + t_1 + (t_3 - t_1)).$$

Then

$$a_2 a_3 x_2^2 x_3^2 = (a_1 x_1^2 + (t_2 - t_1))(a_1 x_1^2 + (t_3 - t_1)).$$

Dividing out by b_i, we get an equation of the form

$$B_2 B_3 (x_2 x_3)^2 = (B_1 x_1^2 + R)(B_1 x_1^2 + R')$$

where

$$B_1 = a_1/b_i,\ B_2 = a_2/b_i,\ B_3 = a_3/b_i,\ R = (t_2 - t_1)/b_i,\ R' = (t_3 - t_1)/b_i.$$

Note that each of the above integers are $\leq 3\delta^2$ in absolute value and choice of δ depends only on ϵ. This is a superelliptic equation. We apply Theorem 1.2.3 for $y^2 = f(x)$ with

$$f(X) = \frac{1}{B_2 B_3}(B_1 X^2 + R)(B_1 X^2 + R') \text{ and } y = x_2 x_3.$$

Note that $f(X)$ has four simple roots and its coefficients depend only on ϵ. So by Theorem 1.2.3, x_1, x_2, x_3 are bounded by a computable number depending on ϵ which gives by (4.7) that k is bounded by such a number say, $k_1(\epsilon)$ and Step 2 is complete.

Step 3. We show that

If $|S_2| \le \epsilon k/4$, *then* k *is bounded by a number depending only on* ϵ.

The proof of this depends on Lemma 1.4.4. Let

$$S_4 = S_1 \backslash S_2.$$

Then $|S_4| \ge \epsilon k/4$. Further

$$a_i > 3k \text{ for } a_i \in S_4.$$

Let us take four elements a_i, a_j, a_{i_1} and $a_{j_1} \in S_4$ which satisfy

$$a_i a_j = a_{i_1} a_{j_1}.$$

We may assume without loss of generality that $x_i x_j > x_{i_1} x_{j_1}$. Consider

$$\Delta = (n + t_i)(n + t_j) - (n + t_{i_1})(n + t_{j_1}).$$

Suppose $\Delta \neq 0$. Then

$$\begin{aligned} |\Delta| &= a_i a_j \left((x_i x_j)^2 - (x_{i_1} x_{j_1})^2\right) \\ &\ge a_i a_j \left((x_i x_j)^2 - (x_i x_j - 1)^2\right) \\ &\ge 2a_i a_j x_i x_j > (a_i a_j)^{1/2}(a_i x_i^2)^{1/2}(a_j x_j^2)^{1/2} \\ &\ge 3kn. \end{aligned}$$

On the other hand,

$$|\Delta| = n(t_i + t_j - t_{i_1} - t_{j_1}) + t_i t_j - t_{i_1} t_{j_1} < 2kn + k^2.$$

Comparing with the above lower bound, we get $n < k$, a contradiction. Hence $\Delta = 0$ giving

$$n(t_i + t_j - t_{i_1} - t_{j_1}) + t_i t_j - t_{i_1} t_{j_1} = 0.$$

Since $n > k^2$, this is possible only when

$$t_i + t_j = t_{i_1} + t_{j_1}, t_i t_j = t_{i_1} t_{j_1}$$

so that

$$t_i = t_{i_1} \text{ or } t_i = t_{j_1}$$

which gives $i = i_1$ or $i = j_1$. Thus S_4 is a set of a_i' s such that all products $a_i a_j$ are distinct. Since $S_4 \subseteq S_1$, each $a_i \le k(\log k)^{1-\epsilon/2}$. Now we apply Lemma 1.4.4 to get

$$|S_4| \ll k(\log k)^{-\epsilon/2}.$$

Since $|S_4| \geq \epsilon k/4$ we find that k is bounded by a number depending only on ϵ say, $k_2(\epsilon)$.

Proof of Theorem 4.3.1 is complete by taking $k_0(\epsilon) > \max(k_1(\epsilon), k_2(\epsilon))$. □

4.4 Proof of Theorem 4.1.2

We first describe the ideas which go into the proof of Theorem 4.1.2. The reader can get a glimpse of the ingenious way, the box principle is used in the proofs of the lemmas. We recall the following factorization of d from Sect. 2.4. Let

$$d = d_1 d_2$$

where d_1 and d_2 are positive divisors of d with $\gcd(d_1, d_2) = 1$ or 2. It is easy to see that we can write d as above in $2^{\omega(d)+1}$ ways. We use the two representations a_i's and A_i's from Sect. 2.2. In the case of consecutive integers we have all the a_i's and A_i's distinct whenever $n > k^2$. When $d > 1$ it is no longer the case. In the first lemma, we use the A_i's and X_i's. We find *enough* X_i 's which are not equal to 1. Then these X_i's are $\gg k \log k$ since they are divisible by primes $> k$. Further we find *enough* number of A_i's which are $\gg k$. This then yields $n + (k-1)d \gg k^3(\log k)^2$. In the second lemma, we use the a_i's to show that there cannot be two distinct pairs of equal a_i's if there exists a suitable factorization of $d = d_1 d_2$. This is used in the third lemma to show that if $2^{\omega(d)+1} \ll k/\log k$ and the number of distinct a_i's is *few* compared to k then $n + (k-1)d \ll k^3$ which contradicts the first lemma. So one of the conditions does not hold. In the fourth lemma, we show that k is bounded if the number of distinct a_i's is large. Hence $2^{\omega(d)+1} \gg k/\log k$ which would prove the theorem. In all the proofs below we take $k > C_0$ where C_0 is an absolute and effectively computable large number so that the inequalities encountered are valid. Let us recall the following notations.

$$S = \{n_1, \ldots, n_k\} = \{n, n+d, \ldots, n+(k-1)d\}.$$

By (2.7), we have

$$n + id = A_i X_i^2,\ P(A_i) \leq k,\ \gcd(A_i, X_i) = 1,\ \gcd(X_i, X_j) = 1 \text{ for } i \neq j.$$

By (2.6) we write

$$n + id = a_i x_i^2,\ 0 \leq i < k,\ a_i - \text{square free and } P(a_i) \leq k.$$

Put

$$T(A) = \{A_0, \ldots, A_{k-1}\},\ T(a) = \{a_0, \ldots, a_{k-1}\}.$$

Lemma 4.4.1 *Suppose (4.2) holds. Then*

$$n + (k-1)d \gg k^3(\log k)^2$$

whenever $k > C_0$ *where* C_0 *is an absolute constant.*

Proof In order to prove the lemma we can assume that

$$d \le k^3 \tag{4.8}$$

since otherwise the lemma is valid for large k. By the fundamental argument of Erdős, there exists a set $S_1 \subset S$ with $|S_1| \ge k - \pi(k)$ such that

$$\prod_{n+id \in S_1} A_i \le (k-1)! \tag{4.9}$$

see (2.2) with $u_i = A_i$.

Step 1. We show that $n + (k-1)d \gg k^2 \log^2 k$. In fact we show that there exists a set of i's of cardinality at least $k/16$ such that $n + id \gg k^2 \log^2 k$ for i in the set. Our starting point is Theorem 3.1.2 by which we know

$$n + (k-1)d > k^2.$$

So if $i \ge k/4$, then

$$n + id \ge n + (k-1)d/4 \ge \frac{1}{4}(n + (k-1)d) \ge k^2/4.$$

Thus

$$n + id \ge k^2/4 \text{ for } k/4 \le i < k.$$

Let

$$I_0 = \{i : k/4 \le i < k\}; \; I_1 = \{i : k/4 \le i < k, X_i = 1\};$$

$$I_2 = I_0 \backslash I_1 = \{i : k/4 \le i < k, X_i \ne 1\}.$$

Note that if $i \in I_1$, then $n + id = A_i$ and if $i \in I_2$, then $X_i > k$ and these X_i's are pairwise distinct. Further we may assume that each X_i is a prime, since otherwise, there is an i such that X_i is divisible by at least two primes $> k$ so that

$$n + (k-1)d \ge X_i^2 > k^4$$

and the assertion of the lemma is true. Next use (4.9), to get

$$\begin{aligned}\left(\frac{k^2}{4}\right)^{|I_1|-\pi(k)} &\le \prod_{\substack{i\in I_1\\ n+id\in S_1}} (n+id)\\ &\le \prod_{\substack{i\in I_1\\ n+id\in S_1}} A_i\\ &\le \prod_{n+id\in S_1} (n+id) \le (k-1)!\end{aligned}$$

This gives

$$|I_1| \le \frac{k\log k}{\log(k^2/4)} + \pi(k).$$

Therefore

$$|I_2| \ge \frac{3k}{4} - \frac{k\log k}{\log(k^2/4)} - \pi(k) \ge \frac{k}{8} \text{ for } k > C_0.$$

In other words, the set of indices i for which $X_i \neq 1$ is large. Now we use the fact that X_i' s are primes. There are at least $k/8$ such primes. By omitting the smallest $k/16$ of such primes, we see that there exists a subset $I_3 \subseteq I_2$ such that

$$|I_3| \ge k/16 \text{ and for } i \in I_3,\ X_i \ge p_{\pi(k)+k/16} > p_{k/16}.$$

Using (1.3), we find that

$$X_i \ge p_{k/16} \ge (k/16)\log(k/16) \gg k\log k \text{ for } i \in I_3$$

which gives

$$n + id \ge k^2\log^2 k \text{ for } i \in I_3. \tag{4.10}$$

This completes Step 1.

Step 2. Let $i_0 \in I_3$. Let $\nu(A_{i_0})$ denote the number of distinct $i \in I_3$ satisfying $A_i = A_{i_0}, i \neq i_0$. Then we claim that

$$\nu(A_{i_0}) \le 2^{\omega(d)+1}.$$

We may assume that $\nu(A_{i_0}) > 1$. Let $i_1 \in I_3$ be such that $A_{i_1} = A_{i_0}$. Then by (4.10),

$$n + i_1 d \ge k^2\log^2 k$$

by Step 1. Now we apply Lemma 2.4.2 to derive that

$$\nu(A_{i_0}) \le 2^{\omega(d)+1}.$$

Step 3. Let

$$I_4 = \{i \in I_3 : A_i \geq k/(2^{\omega(d)+6})\},$$

so that

$$I_4 = \{i : k/4 \leq i < k, X_i \geq p_{k/16} \text{ and } A_i \geq k/(2^{\omega(d)+6})\}.$$

We show that

$$|I_4| \geq k/32.$$

Let

$$I_5 = I_3 \backslash I_4 = \{i : k/4 \leq i < k, X_i \geq p_{k/16} \text{ and } A_i < k/(2^{\omega(d)+6})\}$$

so that

$$|I_4| + |I_5| \geq k/16$$

since $|I_3| \geq k/16$. Let $i \in I_5$. By **Step** 2, there can be at most $2^{\omega(d)+1}$ number of A_j's which can be equal to A_i. There are at most $k/2^{\omega(d)+6}$ such distinct A_j's. Thus

$$|I_5| \leq k(2^{\omega(d)+1})/(2^{\omega(d)+6}) = k/32.$$

This shows that

$$|I_4| \geq k/32$$

as required.

Step 4. From index sets, now we pass on to the set of A_i's. Let us denote by

$$T_1 = \{A_i : i \in I_4\}$$

and

$$T_2 = \{A_i \in T_1, \nu(A_i) \geq 2\}.$$

Suppose $|T_2| \leq k(64 \times 2^{\omega(d)+1})^{-1}$. Then we show that the lemma holds. For this, note that by **Steps** 2 and 3, we have

$$k/32 \leq |I_4| \leq |T_1| + 2^{\omega(d)+1}|T_2| \leq |T_1| + k/64$$

giving

$$|T_1| \geq k/64.$$

Observe that

$$X_i \gg k \log k \text{ if } A_i \in T_1.$$

Since there are at least $k/64$ distinct A_i's we find that

$$n + (k-1)d \gg (k/64)(k \log k)^2 \gg k^3 \log^2 k$$

which proves the lemma.

Step 5. Now we take

$$|T_2| > k(64 \times 2^{\omega(d)+1})^{-1}. \tag{4.11}$$

We show that this is not possible. Let $A_i \in T_2$. Then there exists $j \neq i$ such that $A_i = A_j$ which gives

$$(i-j)d = A_i(X_i - X_j)(X_i + X_j).$$

Hence there exists a factorization $d = d_1 d_2$ with $\gcd(d_1, d_2) = 1$ or 2 such that

$$d_1 | (X_i - X_j) \text{ and } d_2 | (X_i + X_j).$$

Since there are $2^{\omega(d)+1}$ such factorization and T_2 satisfies (4.11), it follows by box principle that there exist at least

$$\lfloor k(8 \times 2^{\omega(d)+1})^{-2} \rfloor$$

distinct $i \in I_4$ such that $A_i \in T_2$ and for each i there is a $j \neq i$ with $0 \leq j < k$ satisfying

$$A_i = A_j, d_1 | (X_i - X_j), d_2 | (X_i + X_j).$$

Further

$$\max\left(\frac{X_i - X_j}{d_1}, \frac{X_i + X_j}{d_2}\right) \leq \frac{X_i^2 - X_j^2}{d} = \frac{i-j}{A_i} \leq 2^{\omega(d)+6} \tag{4.12}$$

since $i \in I_4$. By (1.4) and the assumption $d < k^3$ by (4.8), we find

$$\omega(d) \ll \frac{\log d}{\log \log d} \ll \frac{\log k}{\log \log k} \ll \log k$$

so that the number of distinct pairs (i, j) is at least

$$\lfloor k(8 \times 2^{\omega(d)+1})^{-2} \rfloor > 2^{\omega(d)+6}.$$

Hence by (4.12) and Box principle there exist distinct pairs (i_1, j_1) and (i_2, j_2) such that

$$\frac{X_{i_1} - X_{j_1}}{d_1} = \frac{X_{i_2} - X_{j_2}}{d_1}, \frac{X_{i_1} + X_{j_1}}{d_2} = \frac{X_{i_2} + X_{j_2}}{d_2}$$

which shows that

$$X_{i_1} = X_{i_2} \text{ and } X_{j_1} = X_{j_2}.$$

This gives $i_1 = i_2$, $j_1 = j_2$, a contradiction. □

As an application of Lemma 4.4.1 we derive some conditions on equal a_i's.

Lemma 4.4.2 *Let (4.2) hold. Let $k > C_0$. Suppose there are a_i, a_j, a_g, a_h satisfying*

$$a_i = a_j, a_g = a_h, i \neq g.$$

Suppose there exists a factorization of $d = d_1 d_2$ with $\gcd(d_1, d_2) = 1$ *or 2 and*

$$d_1 \big| (x_i - x_j), d_1 \big| (x_g - x_h), d_2 \big| (x_i + x_j), d_2 \big| (x_g + x_h).$$

Then

$$a_i = a_g, |x_i - x_j| = |x_g - x_h| \text{ or } a_i = a_g, x_i + x_j = x_g + x_h.$$

Proof Without loss of generality we assume that $x_i > x_j$ and $x_g > x_h$. Then $i > j$ and $g > h$. By the given hypothesis, we have

$$(i - j)d = a_i(x_i - x_j)(x_i + x_j); \; (g - h)d = a_g(x_g - x_h)(x_g + x_h).$$

Consider

$$a_i x_i^2 - a_g x_g^2 = a_i \left(\frac{x_i - x_j}{2} + \frac{x_i + x_j}{2} \right)^2 - a_g \left(\frac{x_g - x_h}{2} + \frac{x_g + x_h}{2} \right)^2$$

which gives

$$4(i - g)d = a_i \left((x_i - x_j) + (x_i + x_j) \right)^2 - a_g \left((x_g - x_h) + (x_g + x_h) \right)^2.$$

Reading the above equality mod d_1 and mod d_2, respectively, and using $\gcd(d_1, d_2) = 1$ *or* 2 we get

$$4 \left(a_i \left(\frac{x_i + x_j}{d_2} \right)^2 - a_g \left(\frac{x_g + x_h}{d_2} \right)^2 \right) \equiv 0 \pmod{d_1}$$

and

$$4 \left(a_i \left(\frac{x_i - x_j}{d_1} \right)^2 - a_g \left(\frac{x_g - x_h}{d_1} \right)^2 \right) \equiv 0 \pmod{d_2}.$$

In other words,

$$4 \left(a_i (x_i + x_j)^2 - a_g (x_g + x_h)^2 \right) \equiv 0 \pmod{dd_2} \tag{4.13}$$

and

$$4\left(a_i(x_i - x_j)^2 - a_g(x_g - x_h)^2\right) \equiv 0 \pmod{dd_1}. \tag{4.14}$$

We consider two cases.

Case 1. Suppose either

$$a_i(x_i + x_j)^2 - a_g(x_g + x_h)^2 = 0$$

or

$$a_i(x_i - x_j)^2 - a_g(x_g - x_h)^2 = 0.$$

Then $a_i = a_g$ since they are square free and this in turn gives either $x_i + x_j = x_g + x_h$ or $x_i - x_j = x_g - x_h$ which gives the assertion of the lemma.

Case 2. Suppose both

$$a_i(x_i + x_j)^2 - a_g(x_g + x_h)^2 \neq 0 \text{ and } a_i(x_i - x_j)^2 - a_g(x_g - x_h)^2 \neq 0.$$

Then from (4.14)

$$dd_1 \leq 4 \max\left(a_i(x_i - x_j)^2, a_g(x_g - x_h)^2\right).$$

We may assume without loss of generality that $a_i(x_i - x_j)^2$ is the maximum. Then

$$dd_1 \leq 4a_i(x_i - x_j)^2.$$

Using $a_i = a_j$ and $x_i > x_j$ we also get

$$n \leq a_i x_j^2 \leq a_i(x_i + x_j)^2/4$$

and so

$$ndd_1 \leq a_i^2(x_i^2 - x_j^2)^2 \leq (a_i x_i^2 - a_j x_j^2)^2 < k^2 d^2.$$

This implies that

$$n < d_2 k^2.$$

Let us use (4.13). Then

$$dd_2 \leq 4\left(a_i x_i^2 - a_g x_g^2 + 2(a_i x_i x_j - a_g x_g x_h) + a_i x_j^2 - a_g x_h^2\right).$$

Note that

$$n \leq a_j x_j^2 < a_i x_i x_j < a_i x_i^2 < n + kd.$$

Similarly $n < a_g x_g x_h < n + kd$ so that $|a_i x_i x_j - a_g x_g x_h| < kd$. Hence $dd_2 \leq 16kd$ giving

$$d_2 \leq 16k.$$

By similar argument with (4.14), we get

$$d_1 \leq 16k.$$

Thus

$$n + (k-1)d \leq d_2 k^2 + k d_1 d_2 \leq 16k^3 + 256k^3 = 272k^3.$$

Since $k > C_0$, Lemma 4.4.1 is valid so that

$$n + (k-1)d \gg k^3 \log^2 k$$

which is not possible for k sufficiently large. □

Lemma 4.4.3 *Let $0 < \epsilon < 1$. Then there exists an effectively computable positive number $C_6(\epsilon)$ such that for $k \geq C_6(\epsilon)$, Eq. (4.2) implies that either*

$$2^{\omega(d)+5} \geq \epsilon \frac{k}{\log k}$$

or

$$|T(a)| \geq k - \epsilon \frac{k}{\log k}.$$

Proof Let $0 < \epsilon < 1$ be given. Let $k > C_0$ so that Lemmas 4.4.1 and 4.4.2 are valid. Let us assume that the assertions of the lemma are false. Then

$$|T(a)| < k - \epsilon \frac{k}{\log k} \tag{4.15}$$

and

$$2^{\omega(d)+5} < \epsilon \frac{k}{\log k}. \tag{4.16}$$

We shall split the proof into three steps.

Step 1. We show that

$$2^{3\omega(d)+8} > \epsilon \frac{k}{\log k}.$$

Suppose not. Then

$$2^{3\omega(d)+8} \leq \epsilon \frac{k}{\log k}. \tag{4.17}$$

Let S_α denote the set of a_i's which are repeated α times and let s_α its cardinality. Note that

$$|T(a)| = \sum_{\alpha \geq 1} s_\alpha \text{ and } k = \sum_{\alpha \geq 1} \alpha s_\alpha$$

so that

$$\sum_{\alpha \geq 2} (\alpha - 1) s_\alpha = k - |T(a)|. \tag{4.18}$$

For $\alpha \geq 2$, we can write

$$S_\alpha = \cup_{i=1}^{s_\alpha} S_{\alpha,i}$$

where each set $S_{\alpha,i}$ has α number of equal a_i's. So, for $\alpha \geq 2$, in $S_{\alpha,i}$, we can find indices

$$t_{1,i} < t_{2,i} < \cdots < t_{\alpha,i}$$

such that $a_{t_{1,i}} = a_{t_{2,i}} = \cdots = a_{t_{\alpha,i}}$. Let

$$V_{\alpha,i} = \{(t_{\alpha,i}, t_{1,i}), \cdots, (t_{\alpha,i}, t_{\alpha-1,i})\}, \ 1 \leq i \leq s_\alpha.$$

Put

$$V_\alpha = \cup_{i=1}^{s_\alpha} V_{\alpha,i} \text{ and } V = \cup_{\alpha \geq 2} V_\alpha.$$

Then by (4.18) and (4.15),

$$|V| = \sum_{\alpha \geq 2} |V_\alpha| = \sum_{\alpha \geq 2} (\alpha - 1) s_\alpha = k - |T(a)| \geq \epsilon \frac{k}{\log k}.$$

Hence by (4.17),

$$|V| \geq 2^{3\omega(d)+8}.$$

So Box principle there exists a set $V' \subset V$ with

$$|V'| \geq 2^{2\omega(d)+7}$$

and a factorization $d = d_1 d_2$ such that

$$d_1 | (x_i - x_j), d_2 | (x_i + x_j) \text{ for } (i, j) \in V'.$$

Fix some a_{i_0}. Suppose $\nu(a_{i_0})$ denotes the number of a_i such that $(i_0, i) \in V'$. Assume that $\nu(a_{i_0}) > 1$. Then there exists i_1 with $(i_0, i_1) \in V'$ such that

$$n + i_1 d \geq \frac{1}{k}(n + (k-1)d) \gg k^2 \log^2 k > k^2$$

for $k > C_0$. Hence by Lemma 2.4.2,

$$\nu(a_{i_0}) \leq 2^{\omega(d)+1}.$$

Since $|V'| \geq 2^{2\omega(d)+7}$, there must exist another pair $(g, h) \in V'$ with $a_{i_0} \neq a_g, a_g = a_h, d_1 | x_g - x_h$ and $d_2 | x_g + x_h$. This contradicts Lemma 4.4.2. Hence

$$2^{3\omega(d)+8} > \epsilon \frac{k}{\log k}$$

which completes Step 1.

Step 2. By Step 1,

$$\omega(d) \gg_\epsilon \log k$$

which together with (1.4) implies that

$$d \gg_\epsilon k^{\log \log k}. \tag{4.19}$$

Step 3. Let us use (4.16). We apply Box Principle as in the previous paragraph to get pairs (i, j) and (g, h), $i \neq g$ in V and a factorization $d = d_1 d_2$ such that

$$a_i = a_j, a_g = a_h, x_i > x_j, x_g > x_h,$$

$$d_1 | (x_i - x_j), d_2 | (x_i + x_j); d_1 | (x_g - x_h), d_2 | (x_g + x_h).$$

Therefore

$$(i - j)d = a_i (x_i^2 - x_j^2)$$

which gives

$$x_i + x_j \leq k d_2. \tag{4.20}$$

Similarly

$$x_g + x_h \leq k d_2.$$

Further by Lemma 4.4.2 we get either

$$a_i = a_g, x_i - x_j = x_g - x_h \text{ or } a_i = a_g, x_i + x_j = x_g + x_h.$$

Let us consider the case

$$a_i = a_g, x_i - x_j = x_g - x_h.$$

The other case is similar. (The reader should check this). Let us take without loss of generality that $i > g$. Then $x_i + x_j \neq x_g + x_h$ and $(x_i - x_j)^2 = (x_g - x_h)^2$ which gives

$$|x_i x_j - x_g x_h| = \frac{1}{2}\left|a_i(x_i^2 - x_g^2) - a_i(x_h^2 - x_j^2)\right| < dk. \tag{4.21}$$

Arguing as in the proof of Lemma 4.4.2 we get from (4.13) with $a_i = a_g$ that

$$4a_i\left((x_i + x_j)^2 - (x_g + x_h)^2\right) \equiv 0 \pmod{dd_2}.$$

Since $\gcd(d, a_i) = 1$, this implies that

$$4\left((x_i + x_j)^2 - (x_g + x_h)^2\right) \equiv 0 \pmod{dd_2} \tag{4.22}$$

which yields by (4.20),

$$d_1 < 4\left(\frac{x_i + x_j}{d_2}\right)^2 < 4k^2. \tag{4.23}$$

Since $x_i - x_j = x_g - x_h$,

$$4\left((x_i - x_j)^2 - (x_g - x_h)^2\right) \equiv 0 \pmod{dd_2}. \tag{4.24}$$

Adding (4.22) and (4.24), we get by (4.21),

$$dd_2 < 16|x_i x_j - x_g x_h| < 16dk$$

giving

$$d_2 < 16k$$

and so by (4.23),

$$d < 64k^3.$$

Comparing this with (4.19) we get a contradiction if $k = k_1(\epsilon)$ is large. Now take $k > \max(k_1(\epsilon), C_0) = C_6(\epsilon)$ to get the assertion of the lemma. □

Proof of Theorem 4.1.2

Step 1. We use the fundamental argument of Erdős as in **Steps** 2 and 3 of Theorem 4.1.1 to show that there exist effectively computable constants c_1 and c_2 such that (4.2) with

$$|T(a)| > k - c_1\frac{k}{\log k}$$

implies that $k \le c_2$. Let us assume that

$$|T(a)| > k - \epsilon \frac{k}{\log k} := K$$

for some $0 < \epsilon < 1$ which we will specify later. By our assumption, we have at least K number of distinct a_i's. Using Step 2 of Theorem 4.1.1, we find

$$\prod_{i=0}^{k-1} a_i \ge 138600 \left(\frac{3}{2}\right)^{K-8} (K-8)!$$

and by Step 1 of Theorem 4.1.1,

$$\prod_{i=0}^{k-1} a_i \le (k-1)!(1.34)^k k^4.$$

Comparing the two bounds, we get

$$5407(1.11)^k < k^{12}(1.5)^{\epsilon k/\log k} e^{\epsilon k}.$$

Taking $\epsilon \le 0.05$ we find that $k < 1540$. Thus we may take $c_1 = 0.05$ and $c_2 = 1540$.

Step 2. Let $k > \max(C_6(\epsilon), 1540)$ with $\epsilon = 0.05$ where $C_6(\epsilon)$ is the quantity appearing in Lemma 4.4.3. By Step 1, we may assume that

$$|T(a)| \le k - (0.05)\frac{k}{\log k}.$$

So Lemma 4.4.3 shows that

$$2^{\omega(d)} \ge \frac{(0.05)k}{32 \log k} = (0.0015)\frac{k}{\log k}$$

proving the assertion of the theorem with

$$C_1 = \max(C_6(0.05), 1540) \text{ and } C_2 = 0.0015.$$

□

4.5 Comments

At present, it is known that (4.2) with $b = 1$ and $4 \le k \le 109$ is not possible. See [12]. In Eqs. (4.1) and (4.2) we have taken $P(b) \le k$. This condition was taken due to 1951 result of Erdős. How far can we relax this condition? In [13], equation (4.2) was considered with $P(b) \le Ck$ where C is fixed but arbitrary. By introducing more combinatorial ideas into the general theory, it was shown that if $\epsilon \in (0, 1)$ and $n \ge (C - d + \epsilon d)k$, then (4.2) implies that

$$\max(n, k, b, y) \le C_1(d, \epsilon, C)$$

where $C_1(d, \epsilon, C)$ is an effectively computable number depending only on d, ϵ and C.

Theorem 4.3.1 shows that the product of terms from k consecutive integers after omitting about $k \log\log k / \log k$ integers cannot be a square if k is large. So one may wonder if we omit few terms, the Eq. (4.5) may have no solution even if k is small. Let us start with omitting one term. Here Erdős and Selfridge observed that

$$\frac{6!}{5} = 12^2, \frac{10!}{7} = 720^2.$$

In [14, Theorem 1], it was shown that these are the only squares that can be written as product of $k - 1$ integers out of $k \ge 3$ consecutive positive integers. As one can expect, if more terms are omitted more exceptions would crop up. This is indeed the case. We refer to papers of Mukhopadhyay and Shorey for many such results, see for instance, Mukhopadhyay and Shorey [15].

Theorem 4.1.2 shows that if $\omega(d)$ is fixed, then k is bounded by an absolute constant. Let us consider $\omega(d) = 1$, i.e., d is a prime. It was proved by Saradha and Shorey [16, Theorem 1] that a product of four or more terms of positive integers in arithmetic progression with common difference a prime power is never a square, i.e., (4.2) *with* $b = 1, k \ge 4$ *and* $d = p^\alpha$, $p -$ *prime*, $\alpha \ge 1$ *has no solution.*

The next improvement came from Laishram and Shorey [17, Theorem 2] when they showed that (4.2) with $b = 1$ and $\omega(d) \le 5$ does not hold. Thus there are infinitely many d values for which (4.2) does not hold. Without any condition on $\omega(d)$ it was also shown that (4.2) with $k \ge 6$ implies that $d > \max\left(10^{10}, k^{\log\log k}\right)$ [17, Theorem 1]. There are several other results spread over many papers. We refer to Shorey's *home page:*http://www.math.tifr.res.in/~ shorey.

Exercise 4

(1) Find all the solutions of $\Delta(n, 3) = 3!y^2$. (This is Lemma 4.2.1).
(2) Is $\Delta(n, 4)$ a square?
(3) Show that a product of consecutive odd integers is never a power.

References

1. O. Rigge, Über ein diophantisches problem, in *9th Congress Mathematicians Scandinavian Helsingfors*, 1938, Mercator (1939), pp. 155–160. [**4**]
2. P. Erdős, Note on the product of consecutive integers (1). J. Lond. Math. Soc. **14**, 194–198 (1939). [**4**]
3. P. Erdős, On a Diophantine equation. J. Lond. Math. Soc. **26**, 176–178 (1951). [**4, 5**]
4. R. Marszalek, On the product of consecutive elements of an arithmetic progression. Monatsch. Math. **100**, 215–222 (1985). [**5**]
5. T.N. Shorey, R. Tijdeman, Perfect powers in products of terms in an arithmetical progression. Compositio Math. **75**, 307–344 (1990). [**4–6**]
6. N. Saradha, On perfect powers in products with terms from arithmetic progressions. Acta Arith. **82**, 147–172 (1997). [**2–5**]
7. L.E. Dickson, *History of the Theory of Numbers*, vol. II (Chelsea, New York, 1952). [**4**]
8. P. Erdős, On the product of consecutive integers III. Indag. Math. **17**, 85–90 (1955). [**1, 4, 5**]
9. T.N. Shorey, Perfect powers in products of integers from a block of consecutive integers. Acta Arith. **49**, 71–79 (1987). [**4**]
10. R. Balasubramanian, T.N. Shorey, Squares in products from a block of consecutive integers. Acta Arith. **65**, 213–220 (1994). [**4**]
11. N. Saradha, T.N. Shorey, Squares in blocks from an arithmetic progression and Galois group of Laguerre polynomials. Int. J. Number Theory **11**, 233–250 (2015). [**4**]
12. N. Hirata-Kohno, S. Laishram, T.N. Shorey, R. Tijdeman, An extension of a Theorem of Euler. Acta Arith. **129**, 71–102 (2007). [**4**]
13. M. Filaseta, S. Laishram, N. Saradha, Solving $n(n+d)\cdots(n+(k-1)d = by^2$ with $P(b) \leq Ck$. Int. J. Number Theory **8**, 161–173 (2012). [**4**]
14. N. Saradha, T.N. Shorey, Almost squares and factorisations in consecutive integers. Compositio Mathematicae **138**, 113–124 (2003). [**4**]
15. A. Mukhopadhyay, T.N. Shorey, Square free part of products of consecutive integers. Publ. Math. Debrecen **64**, 79–99 (2004). [**4**]
16. N. Saradha, T.N. Shorey, Almost squares in arithmetic progression. Compositio Mathematicae **138**, 73–111 (2003). [**1, 4**]
17. S. Laishram, T.N. Shorey, The equation $n(n+d)\cdots(n+(k-1)d) = by^2$ with $\omega(d) \leq 6$ or $d \leq 10^{10}$. Acta Arith. **129**, 249–305 (2007). [**3, 4**]

Chapter 5
Erdős-Selfridge Theorem and Some Generalizations

$$9 \times 18 \times 27 \times 36 = 54^3$$

In this chapter, we deal with cubes and higher powers in products of consecutive integers and products with terms from an arithmetic progression. Consider

$$\Delta(n,k) = by^{\ell} \text{ in integers } n > 0, k > 1, y > 1, b \geq 1, P(b) \leq k, \ell \geq 3, \ell - \text{prime.} \tag{5.1}$$

In 1939, soon after proving the square case, Rigge (not published) and Erdős showed using the Thue-Siegel method that (5.1) with $b = 1$ implies that

$$k \leq k_0(\ell)$$

where $k_0(\ell)$ is an integer depending on ℓ. In 1940, Siegel and Erdős showed by the same method that $k \leq k_0$ where now k_0 is an absolute constant. Again this work was not published. Later in 1955, Erdős [1] showed that this result can be proved without the Thue-Siegel method. See also [2]. In 1966, Erdős joined hands with Selfridge in the University of Illinois at Urbana to work on the above equation. In 1975 they [3], succeeded in improving Erdős method, used ideas from graph theory and computations to completely solve the 150 year old conjecture that the product of two or more consecutive positive integers is never a perfect power, i.e.,

Theorem (Erdős-Selfridge) *Equation* (5.1) *with* $b = 1$ *has no solution.*

In this chapter, we will prove the remarkable result of Erdős and Selfridge as a consequence of a more general result (Theorem 5.1.2). A variation of this equation is considered in Sect. 5.3. For products in arithmetic progression, Erdős *conjectured* that the equation

$$\Delta(n,d,k) = by^{\ell} \text{ in integers } n > 0, d > 1, k > 1, y > 1, b \geq 1, \ell \geq 3, \ell - \text{prime} \tag{5.2}$$

S. Natarajan, *Perfect Powers—An Ode to Erdős*, Infosys Science Foundation Series,
https://doi.org/10.1007/978-981-96-2599-4_5

with $P(b) \le k$ implies that k is bounded by an absolute constant. This conjecture is still *not solved.* Henceforth we shall call this conjecture as Erdős conjecture. What is the significance of this conjecture? Suppose we show that (5.2) implies that $k \le k_0$ where k_0 is an explicitly given absolute constant. Then it means, for $k > k_0$ the equation is solved in all the variables n, d, y, b and ℓ. Marszalek [4] showed that (5.2) with $b = 1$ implies that

$$k \le \begin{cases} \max\left(c, \frac{3}{2}e^{d(d+2)(d+1)^{1/3}/2}\right) \text{ if } \ell = 3 \\ \max\left(c, \frac{3}{2}(d+1)\right) \text{ if } \ell \ge 5 \end{cases}$$

where $c = 3 \times 10^4$. Thus whenever d is bounded, then Erdős' conjecture follows. Shorey and Tijdeman [5, Theorem 1(b)] proved a result for (5.2) analogous to Theorem 4.1.2. We shall prove their result in Sect. 5.3 (Theorem 5.1.3). The reader will once again witness an ingenious application of box principle.

5.1 Statement of the Results

Theorem 5.1.1 *Equation* (5.1) *with* $k \ge 4$ *and* $P(b) \le k$ *has no solution provided* $\Delta(n, k)$ *is divisible by a prime* $> k$.

In fact, we show a stronger result.

Theorem 5.1.2 *Suppose* (5.1) *holds with* $k \ge 4$ *and* $P(b) \le k$. *Further assume that* $\Delta(n, k)$ *is divisible by a prime* $> k$. *Then there is a prime* $p > k$ *for which*

$$\nu_p(\Delta(n, k)) \not\equiv 0 \pmod{\ell}.$$

The above theorems with this formulation appeared in Saradha [6, Theorem 1].

Remarks.

(1) Let $P(b) \le k/2$. Suppose $n \le k$. Then $n + k \ge 2n$. So by Bertrand's postulate, there is a prime between $\lceil \frac{n+k}{2} \rceil$ and $n + k$ and this prime divides the product $\Delta(n, k)$ only to the first power. So (5.1) does not hold. Hence in proving Theorem 5.1.2 with $P(b) \le k/2$, it suffices to assume that $n > k$. In this case, by the Theorem of Sylvester, $\Delta(n, k)$ is divisible by a prime $> k$. In fact, Erdős and Selfridge conjectured that whenever $b = 1$ and $n > k$, then there exists a prime $> k$ which divides $\Delta(n, k)$ to the *first* power. This is still *open.* One can make similar conjecture even if $P(b) \le k/2$.

(2) Let $b = 1$. Suppose (5.1) holds with $k = 2$ or 3. Then by the representations of n_i's and gcd conditions, we have

$$(a_0, a_1) = (1, 1) \text{ if } k = 2$$

and if $k = 3$,

$$(a_0, a_1, a_2) = \begin{cases} (1, 1, 1) \\ (2, 1, 2^{\ell-1}) \\ (2^{\ell-1}, 1, 2). \end{cases}$$

Thus when $k = 2, x_1^\ell - x_2^\ell = 1$ which is not possible. Similarly $k = 3$ and $(a_0, a_1, a_2) = (1, 1, 1)$ is not possible. In the remaining two cases, note that

$$(n+1)^2 - 1 = n(n+2) = (2x_0x_2)^\ell, n + 1 = x_1^\ell$$

which is again not possible. Now the Theorem of Erdős and Selfridge follows from Theorem 5.1.1.

(3) Suppose the hypothesis of Theorem 5.1.1 holds. Then by Theorem 5.1.2, there exists a prime $p > k$ which does not occur to an ℓth power in $\Delta(n, k)$. This contradicts the Eq. (5.1). Thus Theorem 5.1.1 is a consequence of Theorem 5.1.2.

(4) In Theorem 5.1.2, there exists a prime $p > k$ which does not occur to an ℓth power in $\Delta(n, k)$. The lower bound for p was improved by Shorey and Tijdeman [7, Theorem 1] for large k.

Theorem 5.1.3 *Equation* (5.2) *implies that*

$$\ell^{\omega(d)} \gg k\frac{\log\log k}{\log k}$$

whenever $k \geq C$ where C is an effectively computable absolute constant.

In fact, in [5], when $\ell = 3$, it was shown that

$$3^{\omega(d)} \gg k\frac{\log\log\log k}{\log k}.$$

The improvement for $\ell = 3$ in Theorem 5.1.3 was shown by Das et al. [8]. Here the upper bound for k depends on ℓ and $\omega(d)$. Thus for fixed ℓ and $\omega(d)$, Erdős conjecture follows. In particular, if ℓ is fixed and d is a prime, the conjecture is true. As in the Remark after Theorem 4.1.2, it follows that

$$d \geq k^{C' \log\log k}$$

where $C' > 0$ is an effectively computable absolute constant.

5.2 Proof of Theorem 5.1.2

The proof uses the techniques explained in Chap. 2.
Stage 1. We bound k. By the given hypothesis, we have $d = 1$, $|S| = k$ and

$$n + (k - 1)d \geq (k + 1)^{\ell}.$$

In Lemma 2.4.4, since $d = 1$, we have $D_1 = D_2 = \theta = \kappa_0 = 1$. Hence the assertion of Lemma 2.4.4 is valid. Thus the set $T(a)$ has Property $P_{\ell'}$ for every $1 \leq \ell' \leq \ell - 1$. Take $\ell' = 2$. We apply Lemma 2.3.1 to get an upper bound for the number of a_i's which are $\leq X$. At the end of Sect. 2.3 we have explained how this helps to get a good lower bound for each a_i's as

$$a_i \geq \alpha(i - \beta).$$

In the Table 5.1, we give the values of m, T for the construction of the set $U, |U^{(0)}|, |V|, \alpha, \beta$. One may use (2.13) to calculate $|U^{(0)}|$. By taking different sets (U, V), the lower bound for a_i is improved. In the last column of the table, we give the least value of i for which one lower bound is better than the previous lower bound. The reader may also refer to [6, Table 1]. Let the elements of $T(a)$ be arranged in the increasing order as

$$a'_1 < a'_2 < \cdots < a'_k.$$

Table 5.1 Values connected to sets U and V

No.	m	T	$\|U^{(0)}\|$	$\|V\|$	α	β	Least i
1.	–	–	–	–	1	0	1
2.	2	16	17	0.3924X+2	2.5484	19	32
3.	3	16	34	0.3424 X+3	2.9205	37	161
4.	6	36	219	0.2431 X+ 10	4.1135	229	700
5.	7	36	262	0.2356 X+16	4.2444	278	1818
6.	8	39	328	0.2279 X+24	4.3878	352	2543
7.	9	42	384	0.2224 X +39	4.4964	423	3292

Thus

$$a'_i \geq \begin{cases} i \text{ for } 1 \leq i \leq 31; \\ 2.5484(i-19) \text{ for } 32 \leq i \leq 160; \\ 2.9205(i-37) \text{ for } 161 \leq i \leq 699; \\ 4.1135(i-229) \text{ for } 700 \leq i \leq 1817; \\ 4.2444(i-278) \text{ for } 1818 \leq i \leq 2542; \\ 4.3878(i-352) \text{ for } 2543 \leq i \leq 3291; \\ 4.4964(i-423) \text{ for } i \geq 3292. \end{cases}$$

Since $T(a)$ has property $P_{\ell'}$ with $\ell' = 1$, $|T(a)| = k$. By the fundamental argument of Erdős, there exists a subset S_1 with $|S_1| \geq k - \pi(k)$ such that

$$\prod_{a'_i \in S_1} a'_i \leq k!$$

Applying the above bounds for a'_i, we get

$$\log\left(31!(2.5484)^{129}\frac{(141)!}{12!}(2.9205)^{539}\frac{662!}{123!}(4.1135)^{1118}\frac{(1588)!}{470!}\times \right. \tag{5.3}$$
$$\left.(4.2444)^{725}\frac{(2264)!}{1539!}(4.3878)^{749}\frac{(2939)!}{(2190)!}(4.4964)^{k-\pi(k)-3291}\frac{(k-\pi(k)-423)!}{(2868)!}\right)$$
$$-\log(k!) \leq \log\left(\prod_{a'_i \in S_1} a'_i\right) - \log(k!) \leq 0.$$

We first use the approximate values from Theorem 1.1.1 and 1.1 for $\pi(k)$ and for factorials from (1.5) and (1.6) to get

$$2347 + \left(k - k/\log k - 1.2762k/\log^2 k\right)(1.5032) +$$
$$\left(k - k/\log k - 1.2762k/\log^2 k - 422.5\right)\log\left(k - k/\log k - 1.2762k/\log^2 k - 423\right)$$
$$-\left(k - k/\log k - 423\right) - \left(k + .5\right)\log k + k - 1/(12k) \leq 0.$$

Observe that the expression on the left-hand side of the above inequality is an increasing function of k. (Check!) For $k = 8536$, this expression is > 0. Hence it is > 0 for all $k \geq 8536$. This contradicts the above inequality. So we conclude that $k < 8536$. Now we use (5.3) with exact values of $\pi(k)$ to conclude that $k \leq 7268$. In [3, p. 296] and [6, pp. 164–166] the bounds obtained were 30000 and 11380, respectively. The reader can play with the above argument and see if the bound for k can still be reduced.

Stage 2. Let $k \leq 7268$. Using (2.9), we check that $f(k, m) \geq f_0(k, m)$ where

$$\begin{cases} f_0(k,2) \geq 4, 4 \leq k \leq 40;\ f_0(k,3) \geq 8, 41 \leq k \leq 112; \\ f_0(k,4) \geq 16, 113 \leq k \leq 310;\ f_0(k,5) \geq 22, 311 \leq k \leq 646; \\ f_0(k,6) \geq 38, 647 \leq k \leq 1180;\ f_0(k,7) \geq 66, 1181 \leq k \leq 1846; \\ f_0(k,8) \geq 115, 1847 \leq k \leq 2728. \end{cases} \tag{5.4}$$

Hence

$$\binom{f_0(k,m)+\ell-2}{\ell-1} > \ell^m \tag{5.5}$$

for $\ell = 3$ and k, m chosen as in (5.4). Now we show by induction that (5.5) is valid for all $\ell > 3$ and k, m as above. As we have taken $m \leq 8$, we find that

$$f_0(k,m) > m + 1 + \frac{3m(m-1)}{2(9-m)}.$$

Since $\ell \geq 3$, we have

$$\begin{aligned} f_0(k,m) + \ell - 1 &> \ell + m + \binom{m}{2}\frac{1}{3}\left(1 + \frac{m}{9} + \frac{m^2}{9^2} + \cdots\right) \\ &> \ell\left(1 + \frac{m}{\ell} + \binom{m}{2}\frac{1}{\ell^2} + \binom{m}{3}\frac{1}{\ell^3} + \cdots\right) \\ &> \ell\left(1 + \frac{1}{\ell}\right)^m, \end{aligned}$$

thereby showing that

$$\binom{f_0(k,m)+\ell-1}{\ell} > (\ell+1)^m$$

by (5.5). Thus (5.5) is valid for all $\ell \geq 3$ and k, m chosen as in (5.4). But this contradicts Lemma 2.2.1 (i). So we conclude that $k \notin [4, 2728]$. For k with $2729 \leq k \leq 7268$ we use the counting function $F_0(k, m_1, m_2)$ given in (2.11). We find that $F(k, m_1, m_2) \geq F_0(k, m_1, m_2)$ where

$$\begin{cases} F_0(k,4,11) \geq 112, 2729 \leq k \leq 6117;\ F_0(k,4,12) \geq 121, 6118 \leq k \leq 6599; \\ F_0(k,5,11) \geq 195, 6600 \leq k \leq 7268. \end{cases} \tag{5.6}$$

We check that

$$\binom{F_0(k,m_1,m_2)+\ell-2}{\ell-1} > \ell^{m_1}\binom{\ell+m_2-1}{\ell-1} \tag{5.7}$$

for $\ell = 3$ and k, m_1, m_2 as given in (5.6). Now we show by induction that (5.7) is valid for all $\ell > 3$ and k, m_1, m_2 as in (5.6). Note that

$$F_0(k, m_1, m_2) > 1 + m_1 + m_2 + \frac{m_1 m_2}{3} + \frac{m_1(m_1 - 1)(m_2 + 3)}{2(9 - m_1)}$$

and so

$$\begin{aligned} F_0(k, m_1, m_2) + \ell - 1 >& (\ell + m_2)\left(1 + \frac{m_1}{\ell} + \binom{m_1}{2}\frac{1}{\ell^2}\left(1 + \frac{m_1}{3^2} + \frac{m_1^2}{3^4} + \cdots\right)\right) \\ >& (\ell + m_2)\left(1 + \frac{1}{\ell}\right)^{m_1} \end{aligned}$$

and hence the inequality (5.7) is valid for all $\ell \geq 3$ and k, m_1, m_2 as chosen in (5.6). This contradicts Lemma 2.2.1(ii). So $k \notin [2729, 7268]$. This completes the proof of Theorem 5.1.2. □

Remark. The argument in Stage 1 is valid for equation (5.2) whenever the condition on D_1 in Lemma 2.4.4 with $\ell' \leq 2$ holds. The argument in Stage 2 holds for (5.2) provided we ensure that a_i's are *distinct*, i.e., condition on D_1 in Lemma 2.4.4 with $\ell' = 1$ holds. Putting together, we have the following corollary.

Corollary 5.2.1 *Suppose* (5.2) *holds. Then*

$$D_1 > \kappa_0 \frac{\theta}{k}(n + (k - 1)d)^{(\ell-2)/\ell}.$$

The expression on the right-hand side of the above inequality may become <1. Then we may be able to conclude only that $D_1 = 1$. In the next chapter we show that indeed $D_1 > 1$.

5.3 A Variation of Equation (5.1)

Analogous to Sect. 4.3, we consider here

$$(n + t_1)(n + t_2) \cdots (n + t_{|S|}) = by^\ell,\ P(b) \leq k,\ \ell \geq 3,\ \ell - \text{prime} \tag{5.8}$$

where

$$S = \{n + t_1, n + t_2, \ldots, n + t_{|S|}\}$$

is a set of $|S| \leq k$ integers belonging to $\{n, n + 1, \ldots, n + k - 1\}$. In the previous section we have taken $|S| = k$. Let us now take $|S| < k$. Suppose

$$k^\alpha < n \leq k^\ell \text{ with } \alpha > 1.$$

Then (5.8) shows that $P(n + t_i) < k$. By the fundamental argument of Erdős, there exists a subset $S_1 \subset S$ with $|S_1| > |S| - \pi(k)$ such that

$$k^{\alpha(|S|-\pi(k))} \leq \prod_{n+t_i \in S_1} (n+t_i) \leq k^k$$

which gives

$$|S| < \alpha^{-1}k + \pi(k).$$

In other words, when $|S| \geq \alpha^{-1}k + \pi(k)$, then (5.8) does not hold. Alternately, we may omit at most

$$k(1-\alpha^{-1}) - \pi(k)$$

terms from $\{n, n+1, \ldots, n+k-1\}$ and still the product of the remaining terms is not a perfect power. Now we consider the case $n > k^\ell$. In [9], Shorey showed the following result.

Theorem 5.3.1 *Suppose* (5.8) *holds with* $\ell \geq 3, n > k^\ell$ *and*

$$|S| \geq \left(\frac{2\ell^2 - 4\ell + 1}{4(\ell^2 - 3\ell + 2)}\right)k.$$

Then k is bounded by an absolute constant.

The above theorem implies for instance that, when $\ell = 3$, we may omit as much as $\lfloor k/8 \rfloor$ terms from $\{n, n+1, \ldots, n+k-1\}$ and still the product of the remaining terms is not a perfect power for large k. In fact, Shorey proved a more general result than Theorem 5.3.1. He showed that (5.8) with $\ell \geq 3$, $n > k^\ell$ and $0 \leq u < 1$ implies that k is bounded by an effectively computable number depending on ϵ whenever

$$t \geq \nu_\ell k$$

where

$$\nu_\ell \geq \frac{1}{2-u} + \epsilon, \nu_\ell \geq \frac{1}{2}\left(1 + \frac{2\ell - 3 + u}{(2\ell - 4 + u)(\ell - 1)}\right), \epsilon > 0.$$

Taking $u = 0$, we get Theorem 5.3.1. We will be content with proving the above theorem to make the exposition somewhat simpler. In the proof of this theorem, the reader will witness the *confluence* of three techniques, viz., k-free integers (Lemma 1.3.1), Linear Forms in Logarithms of algebraic numbers (Lemmas 1.2.1 and 1.2.2) and rational approximation to algebraic numbers (Lemma 1.2.4).

Proof of Theorem 5.3.1 Let $T(A) = \{A_1, \ldots, A_{|S|}\}$. Since $n > k^\ell$,

$$|T(A)| = |S| \geq \left(\frac{2\ell^2 - 4\ell + 1}{4(\ell^2 - 3\ell + 2)}\right)k.$$

We will take

$$0 < \epsilon < \frac{1}{100\ell^4}. \tag{5.9}$$

Step 1. As usual, our first step is an application of fundamental argument of Erdős. By this argument there exists a subset $S_1 \subset T(A)$, $|S_1| \geq |T(A)| - \pi(k)$ and

$$\prod_{A_i \in S_1} A_i \leq k^k. \tag{5.10}$$

Let

$$S_1' = \left\{ A_i \in S_1 : A_i \geq k^{\frac{4(\ell^2-3\ell+2)(1+\epsilon)}{2\ell^2-4\ell+1}} \right\}$$

where we choose $\epsilon > 0$ so that

$$\frac{4(\ell^2 - 3\ell + 2)(1+\epsilon)}{2\ell^2 - 4\ell + 1} < 2. \tag{5.11}$$

Thus we need

$$\epsilon < \frac{2\ell - 3}{2(\ell^2 - 3\ell + 2)}$$

which is satisfied by (5.9). From now on we shall take $k \geq k_0 = k_0(\ell, \epsilon)$ where k_0 is a large number which depends only on ℓ and ϵ with ℓ sufficiently large. By (5.10), we get

$$k^{\frac{4(\ell^2-3\ell+2)(1+\epsilon)}{2\ell^2-4\ell+1}|S_1'|} \leq k^k$$

which gives

$$|S_1'| \leq \frac{2\ell^2 - 4\ell + 1}{4(\ell^2 - 3\ell + 2)(1+\epsilon)} k.$$

Let

$$S_2 = S_1 \backslash S_1'$$

so that by Theorem 1.1.1

$$\begin{aligned} |S_2| &\geq \frac{2\ell^2 - 4\ell + 1}{4(\ell^2 - 3\ell + 2)} k - \pi(k) - \frac{2\ell^2 - 4\ell + 1}{4(\ell^2 - 3\ell + 2)(1+\epsilon)} k \\ &\geq \frac{2\ell^2 - 4\ell + 1}{4(\ell^2 - 3\ell + 2)} k - \frac{k}{\log k} - \frac{1.2762k}{\log^2 k} - \frac{2\ell^2 - 4\ell + 1}{4(\ell^2 - 3\ell + 2)(1+\epsilon)} k \\ &= \frac{2\ell^2 - 4\ell + 1}{4(\ell^2 - 3\ell + 2)} \frac{\epsilon k}{1+\epsilon} - \epsilon k \end{aligned}$$

by taking $k > e^{2/\epsilon}$. Thus

$$S_2 = \left\{ A_i \in S_1 : A_i < k^{\frac{4(\ell^2-3\ell+2)(1+\epsilon)}{2\ell^2-4\ell+1}} \right\}$$

has

$$|S_2| \geq \frac{2\ell^2 - 4\ell + 1}{4(\ell^2 - 3\ell + 2)} \frac{\epsilon k}{1+\epsilon} - \epsilon k. \tag{5.12}$$

Step 2. Application of Lemma 1.3.1. Take an $A_i \in S_2$. Then

$$k^{\frac{4(\ell^2-3\ell+2)(1+\epsilon)}{2\ell^2-4\ell+1}} X_i^\ell > A_i X_i^\ell > k^\ell$$

since $n > k^\ell$. Thus

$$X_i^\ell > k^{\ell - \frac{4(\ell^2-3\ell+2)(1+\epsilon)}{2\ell^2-4\ell+1}} > k^{\ell-2} > k$$

by (5.11) and $\ell \geq 3$. So $X_i > k$. We have seen on earlier occasions how to improve this to $k \log k$. We will show that there exists a subset $S_3 \subset S_2$ with

$$|S_3| \gg k^{1-\epsilon} \text{ and } X_i > k^{2-\frac{2\ell^2-8\ell+7}{(\ell-1)(2\ell^2-4\ell+1)}-\epsilon} > k^{2-\frac{1}{\ell}-\epsilon} \text{ for } A_i \in S_3. \tag{5.13}$$

We proceed as follows. Let

$$S_4 = \{A_i \in S_2 : X_i \text{ is prime}\} \text{ and } S_5 = S_2 \backslash S_4.$$

Suppose X_i were composite, then we know that $X_i \geq k^2$. Hence if $|S_5| \gg k^{1-\epsilon}$ we are through by taking $S_3 = S_5$. So we may assume that $|S_5| \ll k^{1-\epsilon}$. Then $|S_4| \gg k^{1-\epsilon}$ by (5.12). As before, by Prime Number Theory, there exists

$$S_6 = \{A_i \in S_4, X_i \gg k \log k\} \text{ with } |S_6| \gg k^{1-\epsilon}.$$

Let us arrange the X_i's with $A_i \in S_6$ in the increasing order and without loss of generality assume that they are

$$X_1 < X_2 < \cdots < X_{|S_6|}.$$

Thus $A_1, \ldots, A_{|S_6|}$ are the elements of S_6. In order to prove (5.13), we may assume that $X_1 < k^2$. In fact, we may assume that the number of X_i's which are $< k^2$ is $\gg k^{1-\epsilon}$. We know that

$$|A_i X_i^\ell - A_j X_j^\ell| < k, \, 1 \leq i, j \leq |S_6|.$$

Since $X_j \gg k \log k$, we have $k = o(X_j)$. Hence

$$|A_i X_i^\ell - A_j X_j^\ell| < \eta X_j, \, 1 \leq i, j \leq |S_6| \tag{5.14}$$

where η is sufficiently small as a function of k. But $k \geq k_0(\ell, \epsilon)$ with ℓ large, so that η is sufficiently small as a function of ℓ. Thus (1.9) in Lemma 1.3.1 is valid. Now we fix $j = 1$ and suppose the inequality

$$A_i(X_i - X_1)^{2\ell-1} \leq k^{\tau}(X_i - X_1)^{2\ell-1} \leq \frac{X_1^{\ell}}{2} \tag{5.15}$$

where

$$\tau = \frac{4(\ell^2 - 3\ell + 2)(1+\epsilon)}{2\ell^2 - 4\ell + 1}$$

holds. Then by Lemma 1.3.1, the number of (A_i, X_i) satisfying (5.14) with $j = 1$ and (5.15) is

$$\ll (A_1 X_1)^{\epsilon/4} < k^{\epsilon}$$

since $A_1 X_1 \leq k^4$. But the number of such X_i's is $\gg k^{1-\epsilon}$. Hence the inequality (5.15) is not valid for some i, say i_2 with $1 < i_2 \leq k^{\epsilon}$. Thus

$$X_{i_2} - X_1 \gg X_1^{\ell/(2\ell-1)} k^{-\tau/(2\ell-1)}.$$

Next we consider the inequality (5.15) with X_1 replaced with X_{i_2} to find $i_2 < i_3 \leq 2k^{\epsilon}$ such that

$$X_{i_3} - X_{i_2} \gg X_{i_2}^{\ell/(2\ell-1)} k^{-\frac{\tau}{(2\ell-1)}}.$$

Proceeding iteratively, we find a sequence of integers,

$$1 < i_2 < i_3 < \cdots < i_R < |S_6| \text{ with } i_j \leq (j-1)k^{\epsilon}, 2 \leq j \leq R \text{ where } R = \lfloor k^{1-\epsilon} \rfloor$$

such that

$$X_{i_j} - X_{i_{j-1}} \gg X_{i_{j-1}}^{\ell/(2\ell-1)} k^{-\frac{\tau}{(2\ell-1)}}.$$

Now we take indices $j = R, \lfloor R/2 \rfloor, \lfloor R/2^2 \rfloor, \ldots, \lfloor R/2^N \rfloor$ for some integer N. Then it follows that

$$X_{i_{\lfloor R/2 \rfloor}} - X_{i_{\lfloor R/2^2 \rfloor}} \gg R k^{-\frac{\tau}{(2\ell-1)}} X_{i_{\lfloor R/2^2 \rfloor}}^{\ell/(2\ell-1)} \gg k^{1-\epsilon-\frac{\tau}{(2\ell-1)}} X_{i_{\lfloor R/2^2 \rfloor}}^{\ell/(2\ell-1)}.$$

In particular,

$$X_{i_{\lfloor R/2 \rfloor}} \gg k^{1-\epsilon-\frac{\tau}{(2\ell-1)}} X_{i_{\lfloor R/2^2 \rfloor}}^{\ell/(2\ell-1)}.$$

Replacing $(R/2, R/2^2)$ with $(R/2^2, R/2^3)$ in the above inequalities we get

$$X_{i_{\lfloor R/2 \rfloor}} \gg X_{i_{\lfloor R/2^3 \rfloor}}^{(\ell/(2\ell-1))^2} k^{\left(1-\epsilon-\frac{\tau}{(2\ell-1)}\right)(1+\ell/(2\ell-1))}.$$

Repeating this process N times yields

$$
\begin{aligned}
X_{i_{\lfloor R/2 \rfloor}} &\gg \left(X_{i_{\lfloor R/2^{N+1} \rfloor}}\right)^{(\ell/(2\ell-1))^N} k^{\left(1-\epsilon-\frac{\tau}{(2\ell-1)}\right)\left(1+\ell/(2\ell-1)+(\ell/(2\ell-1))^2+\cdots+(\ell/(2\ell-1))^{N-1}\right)} \\
&\gg k^{\left(1-\epsilon-\frac{\tau}{(2\ell-1)}\right)\left(1-(\ell/(2\ell-1))^N\right)\left(\frac{2\ell-1}{\ell-1}\right)}.
\end{aligned}
$$

Now choose N so that $\left(\frac{\ell}{2\ell-1}\right)^N < \epsilon$. Thus for large ℓ, we have

$$X_{i_{\lfloor R/2 \rfloor}} \gg k^{2+\frac{1}{\ell-1}-\frac{\tau}{\ell-1}-5\epsilon}.$$

Hence

$$X_j \gg k^{2-\frac{2\ell^2-8\ell+7}{(\ell-1)(2\ell^2-4\ell+1)}-5\epsilon} \text{ for } i_{\lfloor R/2 \rfloor} \le j \le R$$

so that at least for $\gg k^{1-\epsilon}$ number of X_i's, the above lower bound is true. Thus S_3 can be taken as

$$S_3 = \{A_i \in S_6 : i_{\lfloor R/2 \rfloor} \le i \le R\}.$$

Step 3. Existence of two A_i's whose ℓth root of their ratio is close to a rational number. Arrange the A_i's in S_3 as

$$A_1 < A_2 < \cdots < A_{|S_3|}.$$

Then

$$
\begin{aligned}
\sum_{i=1}^{|S_3|-1} \log\left(\frac{A_{j+1}}{A_j}\right) &\le \log A_{|S_3|} \\
&\le \frac{4(\ell^2-3\ell+2)(1+\epsilon)}{2\ell^2-4\ell+1} \log k \\
&< 2\log k.
\end{aligned}
$$

So there exists $1 \le i \le |S_3|$ such that

$$\log\left(\frac{A_{i+1}}{A_i}\right) \le \frac{2\log k}{k^{1-\epsilon}} < \log\frac{8}{7}, \tag{5.16}$$

for $k \ge k_0$. Thus

$$A_i \ge \frac{7}{8} A_{i+1}.$$

Consider

$$\left|A_{i+1}X_{i+1}^{\ell} - A_i X_i^{\ell}\right| < k.$$

This gives

$$\left|\frac{A_{i+1}}{A_i} - \frac{X_i^{\ell}}{X_{i+1}^{\ell}}\right| < \frac{k}{A_i X_{i+1}^{\ell}} < \frac{8k}{7A_{i+1}X_{i+1}^{\ell}}.$$

So

$$\left|\left(\frac{A_{i+1}}{A_i}\right)^{1/\ell} - \frac{X_i}{X_{i+1}}\right| < \frac{k}{A_i X_{i+1}^{\ell}} < \frac{8k}{7A_{i+1}X_{i+1}^{\ell}}.$$

Step 4. Application of Baker's Theorem 1.2.4. We use (A_i, X_i) and (A_{i+1}, X_{i+1}) found in **Step** 3. In the notation of Theorem 1.2.4, let us take

$$a = \ell A_{i+1}, b = \ell A_i, m = 1.$$

Since ℓ is a prime,

$$1 < \mu_\ell = \ell^{1/(\ell-1)} < \ell.$$

Since $A_i \in S_2$, we have $A_i < k^{\frac{4(\ell^2-3\ell+2)(1+\epsilon)}{2\ell^2-4\ell+1}}$. By (5.16), we get

$$\frac{7}{8}a = \frac{7}{8}\ell A_{i+1} \le b = \ell A_i \le \ell A_{i+1} = a$$

and

$$\frac{a-b}{b} = \frac{A_{i+1}-A_i}{A_i} \le e^{2\log k/k^{1-\epsilon}} - 1 \le \frac{4\log k}{k^{1-\epsilon}} \text{ for large } k.$$

So

$$\begin{aligned}\lambda = \frac{4b}{(a-b)^2\mu_\ell} &= \frac{4}{\ell^{\ell/(\ell-1)}}\left(\frac{A_{i+1}-A_i}{A_i}\right)^{-2} A_i^{-1} \\ &\ge \frac{1}{4\ell^{\ell/(\ell-1)}\log^2 k} k^{2-2\epsilon-\frac{4(\ell^2-3\ell+2)(1+\epsilon)}{2\ell^2-4\ell+1}} \\ &\ge k^{\frac{4\ell-6}{2\ell^2-4\ell+1}-5\epsilon}\end{aligned}$$

by taking $k > k_0$. Next,

$$2\mu_\ell(a+b) \le 4\ell^2 k^{\frac{4(\ell^2-3\ell+2)(1+\epsilon)}{2\ell^2-4\ell+1}} \le k^{\frac{4(\ell^2-3\ell+2)(1+\epsilon)}{2\ell^2-4\ell+1}+\epsilon}.$$

Hence

$$\begin{aligned}\kappa = 1 + \frac{\log(2\mu_\ell(a+b))}{\log\lambda} &\le 1 + \frac{\frac{4(\ell^2-3\ell+2)(1+\epsilon)}{2\ell^2-4\ell+1}+\epsilon}{\frac{4\ell-6}{2\ell^2-4\ell+1}-5\epsilon} \\ \le 1 + \left(\frac{4(\ell^2-3\ell+2)(1+\epsilon)}{4\ell-6} + \frac{\epsilon(2\ell^2-4\ell+1)}{4\ell-6}\right)&\left(1+\frac{10\epsilon(2\ell^2-4\ell+1)}{4\ell-6}\right) \\ \le 1 + \frac{2(\ell^2-3\ell+2)}{2\ell-3} + \epsilon&\left(\frac{3\ell^2}{(2\ell-3)} + \frac{20\ell^4}{(2\ell-3)^2} + \frac{30\ell^4\epsilon}{(2\ell-3)^2}\right)\end{aligned}$$

$$\leq \ell - \frac{\ell-1}{2\ell-3} + 25\ell^2\epsilon$$
$$\leq \ell - \frac{\ell-1}{2\ell-3} + \frac{1}{4\ell^2}$$

by our assumption on ϵ in (5.9). Then

$$\ell - \kappa \geq \frac{4\ell^3 - 4\ell^2 - 2\ell + 3}{4\ell^2(2\ell-3)}.$$

Lastly we calculate

$$c^{-1} = 2^{\kappa+2}(a+b) \leq 2^\ell 8\ell A_{i+1}.$$

Now we apply Theorem 1.2.4 to find

$$\frac{1}{2^\ell 8\ell A_{i+1} X_{i+1}^{\kappa}} \leq \left| \left(\frac{A_{i+1}}{A_i}\right)^{1/\ell} - \frac{X_i}{X_{i+1}} \right| < \frac{8k}{7A_{i+1}X_{i+1}^{\ell}}$$

by Step 3. So by (5.13) and taking k such that $k^\epsilon > 64\ell 2^\ell$, we get

$$k^{\left(2-\frac{2\ell^2-8\ell+7}{(\ell-1)(2\ell^2-4\ell+1)}-\epsilon\right)(\ell-\kappa)} \leq X_{i+1}^{\ell-\kappa} < 64\ell 2^\ell k < k^{1+\epsilon}.$$

Thus

$$\left(2 - \frac{1}{\ell} - \epsilon\right)\left(\frac{4\ell^3 - 4\ell^2 - 2\ell + 3}{4\ell^2(2\ell-3)}\right) < 1 + \epsilon$$

which simplifies to

$$\epsilon > \frac{8\ell - 3}{12\ell^4}.$$

This contradicts our choice of ϵ in (5.9). Thus $k \leq k_0(\ell)$.

It remains to show that ℓ is bounded by an absolute constant. The proof of this depends on application of Lemmas 1.2.1 and 1.2.2. Shorey [10, Lemma 6] proved this result with $|S| \geq k\ell^{-1/11} + \pi(k) + 2$. This is the first time linear forms in logarithms are applied. We shall state it as a proposition.

Proposition 5.3.2 *Suppose* (5.8) *holds with* $\ell \geq 3$, $n > k^\ell$ *and*

$$|S| \geq k\ell^{-1/11} + \pi(k) + 2.$$

Then there exist absolute constants k_1 *and* ℓ_0 *such that*

$$\ell \leq \ell_0 \text{ whenever } k \geq k_1.$$

Remark. Note that when $|S| \geq \left(\frac{2\ell^2-4\ell+1}{4(\ell^2-3\ell+2)}\right)k$, the hypothesis of the proposition is satisfied for large ℓ, for instance, $\ell > 4^{11}$ and $k \geq k_2$ for some large constant k_2. So the proposition is applicable in the above proof of Theorem 5.3.1. Put $\ell_1 = \max(\ell_0, 4^{11})$. Thus $k \leq \max(k_0(\ell_1), k_1, k_2)$ which completes the proof of Theorem 5.3.1.

Proof of Proposition 5.3.2. We may assume that $k \geq k_1$ where k_1 is a large absolute constant so that all the inequalities are valid where the largeness of k is required. Recall that

$$|T(A)| = |S| \geq k\ell^{-1/11} + \pi(k) + 2$$

and $S_1 \subset T(A)$ in Step 1 of the proof of Theorem 5.3.1 has $|S_1| \geq |T(A)| - \pi(k) \geq k\ell^{-1/11} + 2$ and (5.10) holds.

Step 1. *Application of Lemma* 1.2.1 *for the first time.* We show that

$$\ell \leq k(\log k)^3. \tag{5.17}$$

We may assume that $\ell > k$. Note that for any $A_i \in S_1$, $X_i > 1$ since otherwise,

$$k^k \leq k^\ell \leq n + id = A_i < k^k,$$

a contradiction. So

$$0 \neq |A_i X_i^\ell - A_j X_j^\ell| < k,$$

i.e.,

$$\left|e^{\log(A_i/A_j)+\ell\log(X_i/X_j)} - 1\right| < \frac{k}{A_j X_j^\ell}.$$

Suppose $|\log(A_i/A_j) + \ell\log(X_i/X_j)| \geq 1/2$, then the above inequality doesn't hold since $X_j > k$. So we may assume that

$$Y = |\log(A_i/A_j) + \ell\log(X_i/X_j)| < 1/2.$$

Then

$$\begin{aligned}
|e^Y - 1| &\geq Y - \left(\frac{Y^2}{2!} + \frac{Y^3}{3!} + \cdots\right) \\
&\geq Y - \frac{Y^2}{2}\left(1 + \frac{Y}{2} + \frac{Y^2}{2^2} + \cdots\right) \\
&\geq Y - \frac{Y^2}{2}\left(1 + \frac{1}{2} + \frac{1}{2^2} + \cdots\right) \\
&\geq Y - Y^2 \geq \frac{1}{2}Y
\end{aligned}$$

so that

$$|\log(A_i/A_j)+\ell\log(X_i/X_j)| < \frac{2k}{A_jX_j^\ell} \le \frac{2k}{n}. \tag{5.18}$$

The left-hand side is a linear form in two logarithms. Here and elsewhere in this proof we will be applying Lemmas 1.2.1 and 1.2.2 for linear forms in two logarithms, i.e., 'n'$=2$ in those lemmas. This should not be confused with n occurring in (5.8). In the notation of Lemma 1.2.1, $\alpha_1 = A_i/A_j, \alpha_2 = X_i/X_j$. Note that the height of A_i/A_j is $\le k^k$ since $A_i, A_j \in S_1$. The height of X_i/X_j is $\le (n+k-1)^{1/\ell} \le (2n)^{1/\ell}$, $b_1 = 1, b_2 = \ell$ so that $B = \ell$. Further

$$\Omega' = k\log k, \Omega = k(\log k)(\log 2n)/\ell.$$

We apply Lemma 1.2.1. Denote by C', C'', C_1', C_2' and C_3' positive and absolute numbers. Then

$$\begin{aligned}|\log(A_i/A_j)+\ell\log(X_i/X_j)| &> \exp(-C'\ell^{-1}k(\log k)^2\log n\log\ell)\\ &= n^{-C'\ell^{-1}k(\log k)^2\log\ell}.\end{aligned} \tag{5.19}$$

From (5.18) and (5.19) and $n > k^\ell$, we get

$$k^{\ell-C'k(\log k)^2\log\ell} \le n^{1-C'\ell^{-1}k(\log k)^2\log\ell} < 2k.$$

Thus

$$\frac{\ell}{\log\ell} \le C''k(\log k)^2$$

giving (5.17).

Step 2. *Application of Lemma* 1.2.1 *second time.* We sharpen (5.17). Let $\delta = 1/11$ so that $|S_1| \ge k\ell^{-\delta}+2$ and let

$$S_2 = \{A_i \in S_1 : A_i \le k^{2\ell^\delta}\} \text{ and } S_2' = S_1\setminus S_2.$$

Then by (5.10), we have

$$k^{2\ell^\delta|S_2'|} \le k^k$$

giving $|S_2'| < k/(2\ell^\delta)$. So

$$|S_2| \ge k/\ell^\delta + 2 - k/(2\ell^\delta) \ge k/(2\ell^\delta).$$

Since $n > k^\ell$ and $\delta = 1/11$, for any $A_i \in S_2$, $X_i > 1$. Further

$$|S_2| \ge k/(2\ell^\delta) \ge 2$$

by (5.17). Take any two elements $A_i, A_j, i \neq j$ from S_2. As in Step 1, we apply Lemma 1.2.1 to

$$|\log(A_i/A_j) + \ell \log(X_i/X_j)| < \frac{2k}{n}$$

with $\alpha_1 = A_i/A_j, \alpha_2 = X_i/X_j$, height of A_i/A_j not exceeding $k^{2\ell^\delta}$, height of X_i/X_j not exceeding $(2n)^{1/\ell}, b_1 = 1, b_2 = \ell$ so that $B = \ell$. Then $\Omega' = 2\ell^\delta \log k, \Omega = 2\ell^{\delta-1}(\log k \log 2n)$ and

$$|\log(A_i/A_j) + \ell \log(X_i/X_j)| > \exp(-C_1'\ell^{\delta-1}(\log \ell)^2(\log n)(\log k) \log\log k).$$

Comparing with the upper bound, we get

$$\ell - C_1'\ell^\delta(\log \ell)^2(\log k) \log\log k < 1$$

giving

$$\ell \leq (\log k)^3. \tag{5.20}$$

Hence

$$|S_2| \gg \frac{k}{(\log k)^{3\delta}}.$$

Step 3. *Application of Lemma* 1.2.2 *in an inductive way to increase the lower bound for n*. Firstly, we use the strategy of Step 3 in the Proof of Theorem 5.3.1. Re-arrange the $n + t_i$'s so that

$$A_1 < \cdots < A_{|S_2|} \text{ are in } S_2.$$

Then by (5.20),

$$\sum_{j=1}^{|S_2|-1} \log\left(\frac{A_{j+1}}{A_j}\right) \leq (\log k)^2.$$

Since $|S_2| \gg \frac{k}{(\log k)^{3\delta}}$, there exists μ with $1 \leq \mu \leq |S_2|$ such that

$$0 < \log\left(\frac{A_{\mu+1}}{A_\mu}\right) \leq \frac{(\log k)^3}{k} \leq k^{-\frac{1}{2}}.$$

As in Step 2, we get a linear form as

$$\left|\log(A_{\mu+1}/A_\mu) + \ell \log(X_{\mu+1}/X_\mu)\right| < \frac{2k}{n}. \tag{5.21}$$

Hence

$$\begin{aligned}|\log(X_{\mu+1}/X_\mu)| &\le \frac{2k}{n} + \log\left(\frac{A_{\mu+1}}{A_\mu}\right)\\ &\le \frac{2}{k^{\ell-1}} + \frac{(\log k)^3}{k}\\ &\le k^{-\frac{1}{2}}.\end{aligned}$$

For the *first time* we apply Lemma 1.2.2 to the linear form in (5.21) with $\alpha_1 = \frac{A_{\mu+1}}{A_\mu}$, $\alpha_2 = \frac{X_{\mu+1}}{X_\mu}$ so that 'A_1'$= k^{2\ell^\delta}$, 'A'$= (2n)^{1/\ell}$, $\tau_1 = 2\ell^\delta$ and $\tau_2 = 4\ell^\delta$. Further $b_1 = 1$ and $b_2 = \ell$. Then $B = \max(b_1, b_2) = \ell < A_1$, $u = 1$ and

$$\exp(-(8C_4\tau_2)^9\tau_1 \log A) > \exp(-C_2'\ell^{-\delta}\log n).$$

Suppose $n^{-C_2'\ell^{-\delta}} \le 2k/n$, then $k^{\ell - C_2'\ell^{1-\delta}} \le 2k$ which shows that ℓ is bounded by an absolute constant. So we may assume that $n^{-C_2'\ell^{-\delta}} > 2k/n$ which by (5.21) implies that

$$|\log(A_{\mu+1}/A_\mu) + \ell\log(X_{\mu+1}/X_\mu)| < \exp(-C_2'\ell^{-\delta}\log n).$$

So by Lemma 1.2.2, we conclude that there exists a prime $q_1 \ll \ell^\delta \le \log k$ by (5.20), an integer j_1 with $0 \le j_1 < q_1$ and a rational number η such that

$$\left(\frac{A_{\mu+1}}{A_\mu}\right)^{j_1}\left(\frac{X_{\mu+1}}{X_\mu}\right) = \eta^{q_1}.$$

By the properties of X_μ's (which properties?) we see that

$$X_\mu = Y_\mu^{q_1},\ X_{\mu+1} = Y_{\mu+1}^{q_1} \text{ for some positive integers } Y_\mu \text{ and } Y_{\mu+1}$$

giving $X_\mu > k^{q_1}$, $X_{\mu+1} > k^{q_1}$ and so

$$n > k^{\ell q_1}.$$

Now we can re-write (5.21) as

$$|\log(A_{\mu+1}/A_\mu) + \ell q_1\log(Y_{\mu+1}/Y_\mu)| < \frac{2k}{n}.$$

We apply Lemma 1.2.2 for the *second time* to the above linear form with $n = 2$, $\alpha_1 = \frac{A_{\mu+1}}{A_\mu}$, $\alpha_2 = \frac{X_{\mu+1}}{X_\mu}$ so that 'A_1'$= k^{2\ell^\delta}$, 'A'$= (2n)^{1/\ell q_1}$, $\tau_1 = 2\ell^\delta$, $\tau_2 = 4\ell^\delta$ and $u = 1$, $B = \ell q_1 < \ell\log k < A_1$ to conclude that Y_μ and $Y_{\mu+1}$ are q_2 th powers for some prime $q_2 \le \log k$. Now we proceed inductively. We carry out $K = \lceil(\log\log k)^2\rceil$ inductive steps. Then there are primes $q_1, \ldots, q_K$ which do not exceed $\log k$ such that if $L = q_1\cdots q_K$, then

$$L = q_1 \dots q_K \geq 2^K > 2^{(\log\log k)^2}, \tag{5.22}$$

$$B = \ell L \leq \ell(\log k)^K = e^{\log \ell (\log\log k)^3} \leq A_1 = k^{2\ell^\delta},$$

$$n + d_\mu = A_\mu Z_\mu^{\ell L}, n + d_{\mu+1} = A_{\mu+1} Z_{\mu+1}^{\ell L},$$

for some positive integers Z_μ, $Z_{\mu+1}$ and

$$n > k^{\ell L}.$$

Again we consider the linear form

$$|\log(A_{\mu+1}/A_\mu) + \ell L \log(Z_{\mu+1}/Z_\mu)| < \frac{2k}{n} < n^{-\frac{1}{2}}.$$

We apply Lemma 1.2.1 with 'A_1'$= k^{2\ell^\delta}$, 'A_2'$= (2n)^{1/\ell L}$, $B = \ell L$ and $\ell \leq (\log k)^3$ by Step 2, to conclude that

$$|\log(A_{\mu+1}/A_\mu) + \ell L \log(Z_{\mu+1}/Z_\mu)| > \exp(-C_3' L^{-1}(\log L)(\log k)(\log\log k)^2(\log n)).$$

Comparing with the upper bound, we get

$$L \ll \log k (\log\log k)^2 \log L$$

or

$$\log L \ll (\log\log k)^\theta \text{ with } \theta < 2$$

which by (5.22) shows that k is bounded by an absolute constant k_3. Then by (5.20),

$$\ell \leq (\log k_3)^3 := \ell_0$$

which completes the proof of the proposition. □

5.4 Proof of Theorem 5.1.3

We assume that (5.2) holds. Let $\epsilon > 0$ be arbitrarily small and $k > k_0(\epsilon)$ where $k_0(\epsilon)$ is very large so that all inequalities are valid where largeness of k is required. Further assume that

$$\ell^{\omega(d)} \leq (1-\epsilon)k \frac{\log\log k}{\log k}.$$

As usual, let $S = \{n, n+d, \dots, n+(k-1)d\}$, $T(A) = \{A_0, A_1 \dots, A_{k-1}\}$ and $T(a) = \{a_0, a_1, \dots, a_{k-1}\}$. By Theorem 3.1.2, we have

$$n + (k-1)d \geq (k+1)^{\ell} \text{ and so } n + d \geq k^{\ell-1}. \tag{5.23}$$

We write

$$\rho = e^{2\pi i/\ell},\ K = \mathbb{Q}(\rho).$$

Let $\mathcal{O}_K$ be the ring of algebraic integers of K and $disc(K)$, the discriminant of K. It is well known that

$$[K : \mathbb{Q}] = \ell - 1,\ |disc(K)| = \ell^{\ell-2}.$$

We use the following basic fact about the prime divisors of $X^{\ell} - Y^{\ell}$ which we have already encountered in Lemma 2.4.4. Let ℓ be a prime as usual. Every prime factor of

$$\frac{X^{\ell} - Y^{\ell}}{X - Y}$$

is either ℓ or $\equiv 1 \pmod{\ell}$. Further ℓ occurs in the factorization of the above expression at most to the first power. By the fundamental argument of Erdős, there exists a subset $S_1 \subset T(A)$ with $|S_1| \geq |T(A)| - \pi(k)$ such that

$$\prod_{A_i \in S_1} A_i \leq k!$$

Step 1. *About equal* A_i*'s.* Suppose $A_i = A_j$ for some A_i and $A_j \in T(A)$. Then we shall show that the ratio $\frac{X_i}{X_j}$ belongs to at most $\ell^{\omega(d)}$ residue classes $\pmod{d}$. We have the identity

$$(j-i)d = A_i(X_j^{\ell} - X_i^{\ell}) = A_i \prod_{h=1}^{\ell}(X_j - \rho^h X_i).$$

Since $\gcd(d, A_i) = 1$, we see that $A_i < k$. Hence by (5.23), we get $X_i > k$, $X_j > k$ and $\gcd(X_i, X_j) = 1$. Further

$$X_i^{\ell} - X_j^{\ell} \equiv 0 \pmod{d}.$$

Hence the ratio X_i/X_j belongs to at most $\ell^{\omega(d)}$ residue classes by Lemma 1.2.5.
Step 2. Let $n + (k-1)d \geq k^{4-\epsilon}$ and $|T(A)| < k - (1-\epsilon/2)k\frac{\log\log k}{\log k}$. We claim that for $k \geq k_0(\epsilon)$, there are at least

$$\left\lceil (1-\epsilon)k\frac{\log\log k}{\log k} \right\rceil$$

distinct pairs (i, j) with X_i/X_j belonging to distinct residue classes. Let

$$I_1 = \left\{ i : k \geq i \geq \epsilon k \frac{\log \log k}{4 \log k} \right\} \text{ and } S_2 = \{A_i \in S_1 : i \in I_1\}.$$

For $A_i \in S_2$, let

$$I_2(A_i) = \{j \in I_1 : A_j = A_i\}$$

and for $h \geq 1$, let

$$S_3(h) = \{A_i : |I_2(A_i)| = h\}.$$

Then clearly

$$\sum_{h \geq 1} h|S_3(h)| = |I_1| \geq k - \epsilon k \frac{\log \log k}{4 \log k} \text{ and } \sum_{h \geq 1} |S_3(h)| = |S_2| \leq |S_1|.$$

Let $|I_2(A_{i_1})| = h$ and $i_1 < i_2 < \cdots < i_h$ be such that $A_{i_1} = A_{i_2} = \cdots = A_{i_h}$. Then there are $\frac{h(h-1)}{2}$ distinct pairs (i, j) with $i > j$ and $A_i = A_j = A_{i_1}$. Hence the number of distinct pairs (i, j) with

$$k > i > j \geq \epsilon k \frac{\log \log k}{4 \log k} \text{ and } A_i = A_j \tag{5.24}$$

is

$$\begin{aligned}
\sum_{h \geq 1} \frac{h(h-1)}{2} |S_3(h)| &\geq \sum_{h \geq 1} h|S_3(h)| - \sum_{h \geq 1} |S_3(h)| \\
&\geq \left(k - \frac{\epsilon k \log \log k}{4 \log k}\right) - |S_1| \\
&\geq k - \frac{\epsilon k \log \log k}{4 \log k} - |T(A)| + \pi(k) \\
&\geq (1 - 3\epsilon/4) k \frac{\log \log k}{\log k}.
\end{aligned}$$

Hence we can find at least

$$\left\lceil (1 - \epsilon) k \frac{\log \log k}{\log k} \right\rceil$$

distinct pairs (i, j) with (5.24). Thus the claim is true if we show that

$$X_{j_1}/X_{i_1} \text{ and } X_{j_2}/X_{i_2}$$

belong to distinct residue classes (mod d) for two distinct pairs (i_1, j_1) and (i_2, j_2) satisfying (5.24). We now proceed to prove this. Suppose there exist pairs $(i_1, j_1) \neq (i_2, j_2)$, with

$$\frac{X_{i_1}}{X_{j_1}} \equiv \frac{X_{i_2}}{X_{j_2}} (\text{mod } d). \tag{5.25}$$

Assume $X_{i_1} X_{j_2} > X_{j_1} X_{i_2}$. Then $X_{i_1} X_{j_2} \geq X_{j_1} X_{i_2} + d$ by (5.25). Consider

$$\begin{aligned} \delta &= (n + i_1 d)(n + j_2 d) - (n + j_1 d)(n + i_2 d) \\ &= A_{i_2} A_{j_1} \left((X_{i_1} X_{j_2})^{\ell} - (X_{j_1} X_{i_2})^{\ell} \right). \end{aligned}$$

Then

$$\begin{aligned} |\delta| &\geq (A_{i_2} A_{j_1})^{1/\ell} \ell d \left((A_{i_2} X_{i_2}^{\ell})(A_{j_1} X_{j_1}^{\ell}) \right)^{(\ell-1)/\ell} \\ &\geq \ell d \left(n + \epsilon k d \frac{\log \log k}{4 \log k} \right)^{2(\ell-1)/\ell} \\ &\geq \ell d \frac{\epsilon^2}{16} \left(\frac{(n + (k-1)d) \log \log k}{\log k} \right)^{2(\ell-1)/\ell}. \end{aligned}$$

On the other hand,

$$|\delta| \leq 2kd(n + (k-1)d).$$

Comparing the lower and upper bounds for $|\delta|$, we get

$$n + (k-1)d < \left(\frac{8k}{\ell \epsilon^2} \right)^{\ell/(\ell-2)} \left(\frac{\log k}{\log \log k} \right)^{2(\ell-1)/(\ell-2)}$$

giving

$$n + (k-1)d < k^{4-\epsilon}$$

since k is sufficiently large. This contradiction proves the claim.

Our next step is to deal with the condition on $n + (k-1)d$. We know that $n + (k-1)d \geq (k+1)^{\ell} > k^{4-\epsilon}$ for $\ell \geq 5$. The only case for which this condition may not be satisfied is $\ell = 3$.

Step 3. Let us consider $\ell = 3$ with $n + (k-1)d < k^{4-\epsilon}$. We arrive at a contradiction. Let

$$I_3 = \{i : 1 \leq i < k \text{ and } X_i = 1\}.$$

For any $i \in I_3$, we have $A_i = n + id \geq n + d > k^2$. By the fundamental argument of Erdős, there exists a subset $I_4 \subset I_3$ with $|I_4| \geq |I_3| - \pi(k)$ and

$$\prod_{i \in I_4} A_i | k!.$$

This implies that

$$k^{2(|I_3| - \pi(k))} \leq (n + d)^{|I_3| - \pi(k)} \leq k^k$$

giving

$$|I_3| \leq k/2 + \pi(k) \leq 2k/3$$

since k is large. Let

$$I_4 = [1, k)\backslash I_3.$$

Then $|I_4| \geq k/3 - 1 \geq k/4$ and for $i \in I_4$, we have $X_i \neq 1$ and hence

$$A_i \leq (n + (k-1)d)/k^3 < k^{1-\epsilon} \text{ for } i \in I_4.$$

So there can be at most $k^{1-\epsilon}$ number of distinct A_i's. Since the total number of A_i's with $i \in I_4$ exceeds $k/4$, we find that there exist at least $k^\epsilon/4$ number of A_i's which are equal. We claim now that the number of equal A_i's does not exceed $9^{\omega(d)} + 1$. Let us first see how this leads to a contradiction. We use (1.4) and $d < k^{3-\epsilon}$ to get for some constant $\eta > 0$ that

$$k^\epsilon/4 \leq 9^{\omega(d)} + 1 < k^{\eta(3-\epsilon)/\log\log k} < k^\epsilon/4$$

for $k > k_0(\epsilon)$ which gives a contradiction. We proceed to show the claim. For any prime $p \neq 3$, we have the following possibilities for the ideal $[p]$ in $K = \mathbb{Q}(\rho)$.

$$[p] = \wp \text{ or } \wp^2 \text{ or } \wp_1\wp_2$$

where $\wp, \wp_1, \wp_2$ are prime ideals in K. Thus putting $d' = d/3^{\nu_3(d)}$ we can write

$$[d'] = D_1 D_2 D_3$$

where D_1, D_2, D_3 are coprime ideals of K in at most $9^{\omega(d')}$ ways. Let $Z = 9^{\omega(d)} + 2$. Suppose there exist indices $1 \leq i_1 < i_2 < \cdots < i_Z < k$ such that

$$A_{i_1} = \cdots = A_{i_Z}.$$

For any i_r with $2 \leq r \leq Z$, we have

$$(i_r - i_1)d = A_{i_1}(X_{i_r}^3 - X_{i_1}^3).$$

Since $\gcd(A_{i_1}, d) = 1$ the ideal $[d]$ divides $P_r = [X_{i_1} - X_{i_r}][X_{i_1} - \rho X_{i_r}][X_{i_1} - \rho^2 X_{i_r}]$ in K. There are $Z - 1 > 9^{\omega(d)}$ products P_r. Hence there exist coprime ideals D_1, D_2, D_3 of K and indices r, s such that $2 < i_r < i_s < Z$ with

$$D_h \big| [X_{i_1} - \rho^h X_{i_r}]; \; D_h \big| [X_{i_1} - \rho^h X_{i_s}], h = 1, 2, 3.$$

So $D_h \big| [X_{i_r} - X_{i_s}]$ implying

$$d' \big| (X_{i_r} - X_{i_s}).$$

Since $A_{i_r} = A_{i_s}$, we have $d \big| (X_{i_r}^3 - X_{i_s}^3)$. Further we know that

$$9 \nmid \frac{X_{i_r}^3 - X_{i_s}^3}{X_{i_r} - X_{i_s}}.$$

Thus we get

$$3^{\mathrm{ord}_3(d)-1} \big| (X_{i_r} - X_{i_s}) \text{ if } 3 \big| d.$$

Putting together the above facts, we find that

$$|X_{i_r} - X_{i_s}| > d/3. \tag{5.26}$$

From the identity

$$(i_r - i_s)d = A_{i_r}(X_{i_r}^3 - X_{i_s}^3)$$

and (5.26) it follows that

$$kd \geq d(A_{i_r} X_{i_r}^3)^{2/3} \geq d \left(\frac{(n + (k-1)d)}{k} \right)^{2/3}$$

which simplifies to

$$n + (k-1)d \leq k^{5/2}.$$

This is a contradiction since $n + (k-1)d > k^3$.

Step 3 shows that $\ell = 3$ with $n + (k-1)d < k^{4-\epsilon}$ does not arise. So even when $\ell = 3$ we may assume that $n + (k-1)d > k^{4-\epsilon}$. We put together Steps 1–3 to get the following lemma.

Lemma 5.4.1 *Let $\epsilon > 0$. Assume that* (5.2) *holds and $k > k_0(\epsilon)$ where $k_0(\epsilon)$ is very large. Suppose*

$$\ell^{\omega(d)} \leq (1-\epsilon)k \frac{\log \log k}{\log k}. \tag{5.27}$$

Then

$$|T(A)| \geq k - (1 - \epsilon/2)k \frac{\log \log k}{\log k}.$$

Step 4. Suppose (5.27) holds. By Lemma 5.4.1, we get

$$|S_1| \geq k - (1-\epsilon/2)k \frac{\log \log k}{\log k} - \pi(k) \geq k - (1 - \epsilon/3)k \frac{\log \log k}{\log k}.$$

Let

$$S_2 = \{A_i \in S_1 : A_i \leq k e^{(1-\epsilon/4) \log \log k}\}.$$

Let us find the cardinality of this set. Suppose $|S_2| < \epsilon k/20$. In the estimate below, we shall use two facts, viz., $n! > (n/e)^n$ for $n \geq 1$ and the function $(y/x)^y$ is a monotonic decreasing function of y when $0 < y < x/e$. So for $k \geq k_0(\epsilon)$,

$$\begin{aligned}
k! &\geq \prod_{A_i \in S_1} A_i \geq (|S_2|)! \left(ke^{(1-\epsilon/4)\log\log k}\right)^{|S_1|-|S_2|} \\
&\geq \left(\frac{|S_2|}{e}\right)^{|S_2|} \frac{\left(ke^{(1-\epsilon/4)\log\log k}\right)^{k\left(1-(1-\epsilon/3)\frac{\log\log k}{\log k}\right)}}{\left(ke^{(1-\epsilon/4)\log\log k}\right)^{|S_2|}} \\
&= \left(\frac{|S_2|}{ek}\right)^{|S_2|} \frac{\left(ke^{(1-\epsilon/4)\log\log k}\right)^{k\left(1-(1-\epsilon/3)\frac{\log\log k}{\log k}\right)}}{\left(e^{(1-\epsilon/4)\log\log k}\right)^{|S_2|}} \\
&\geq \left(\frac{\epsilon}{20e}\right)^{\epsilon k/20} \frac{k^k e^{(1-\epsilon/4)k\log\log k}}{e^{(1-\epsilon/3)k\log\log k+(1-\epsilon/4)(1-\epsilon/3)k\frac{(\log\log k)^2}{\log k}+\epsilon(1-\epsilon/4)k\log\log k/20}} \\
&\geq k^k e^{(\epsilon/12-\epsilon/20-\epsilon/60)k\log\log k} \\
&> k^k,
\end{aligned}$$

a contradiction. Thus $|S_2| \geq \epsilon k/20$. Further

$$X_i > k \text{ if } A_i \in S_2$$

and since there are at least $\epsilon k/20$ such X_i's we get

$$n + (k-1)d \gg k^{\ell+1}. \tag{5.28}$$

Step 5. *Application of Lemma* 1.4.4. Define

$$S_3 = \{A_i \in S_2 : i \geq \epsilon k/80 \text{ and } A_i \geq \epsilon k/80\}.$$

Then

$$|S_3| \geq |S_2| - 2\epsilon k/80 \geq \epsilon k/40.$$

Let S_4 be the maximal subset of S_3 such that all products $A_i A_j$ with $A_i, A_j \in S_3$ are distinct. Then by Lemma 1.4.4 and the fact that A_i's are in S_2, we get

$$|S_4| \leq \frac{2ke^{(1-\epsilon/4)\log\log k}}{\log k} = \frac{2k}{(\log k)^{\epsilon/4}}.$$

Let $S_5 = S_3 \backslash S_4$ so that

$$|S_5| \geq \frac{\epsilon k}{40} - \frac{2k}{(\log k)^{\epsilon/4}} \geq \frac{\epsilon k}{41}.$$

Let $A_i \in S_5$. Then there exist $A_{i_1}, A_{i_2}, A_{i_3} \in S_4$ such that

$$A_i A_{i_1} = A_{i_2} A_{i_3}.$$

Therefore,

$$(X_i X_{i_1})^\ell - (X_{i_2} X_{i_3})^\ell \equiv 0 \pmod{d}.$$

Also note that $X_i X_{i_1} - X_{i_2} X_{i_3} \neq 0$. (Why?) Let $R(\ell, d)$ denote the number of residue classes $z \pmod{d}$ such that $z^\ell \equiv 1 \pmod{d}$. By Lemma 1.2.5, $R(\ell, d) \leq \ell^{\omega(d)}$. Hence the ratio $X_i X_{i_1}/X_{i_2} X_{i_3}$ belongs to at most $\ell^{\omega(d)}$ residue classes $\pmod{d}$. By our assumption (5.27), the ratio $X_i X_{i_1}/X_{i_2} X_{i_3}$ belongs to $\ll k \frac{\log\log k}{\log k}$ residue classes. On the other hand, $A_i \in S_5$ and $|S_5| \geq \epsilon k/41$. So we derive that there exist two elements A_i and $A_j \in S_5$ and two triples $A_{i_1}, A_{i_2}, A_{i_3}$ and $A_{j_1}, A_{j_2}, A_{j_3}$ from S_4 such that

$$A_i A_{i_1} = A_{i_2} A_{i_3} \text{ and } A_j A_{j_1} = A_{j_2} A_{j_3}$$

and

$$\frac{X_i X_{i_1}}{X_{i_2} X_{i_3}} \equiv \frac{X_j X_{j_1}}{X_{j_2} X_{j_3}} \pmod{d}.$$

Let us put

$$\xi = X_i X_{i_1} X_{j_2} X_{j_3} - X_j X_{j_1} X_{i_2} X_{i_3}.$$

Then $\xi \neq 0$ and $d | \xi$. Without loss of generality, let us assume that $\xi > 0$. Note that

$$A_i A_{i_1} A_{j_2} A_{j_3} = A_j A_{j_1} A_{i_2} A_{i_3}.$$

Consider

$$\delta = (n + id)(n + i_1 d)(n + j_2 d)(n + j_3 d) - (n + jd)(n + j_1 d)(n + i_2 d)(n + i_3 d).$$

Note that $\delta > 0$. Since A_i's belong to S_3 they are all $\gg k$ and each $n + hd \gg (n + (k-1)d, h \in \{i, i_1, i_2, i_3, j, j_1, j_2, j_3\}$. We use these facts below. On one hand, we have

$$\begin{aligned}
\delta &= A_j A_{j_1} A_{i_2} A_{i_3} \left((X_i X_{i_1} X_{j_2} X_{j_3})^\ell - (X_j X_{j_1} X_{i_2} X_{i_3})^\ell\right) \\
&\geq \ell (A_j A_{j_1} A_{i_2} A_{i_3})^{1/\ell} \xi \left(A_j X_j^\ell A_{j_1} X_{j_1}^\ell A_{i_2} X_{i_2}^\ell A_{i_3} X_{i_3}^\ell\right)^{(\ell-1)/\ell} \\
&\gg \ell k^{4/\ell} \xi (n + (k-1)d)^{4(\ell-1)/\ell} \\
&\gg \ell k^{4/\ell} d (n + (k-1)d)^{4(\ell-1)/\ell}.
\end{aligned}$$

On the other hand,

$$\delta \leq (n + (k-1)d)^4 - n^4 \leq 4kd(n + (k-1)d)^3.$$

Comparing the two bounds of δ, we get

$$\ell(n+(k-1)d)^{1-4/\ell} \ll k^{1-4/\ell}$$

which is not possible for $\ell \geq 5$ by (5.28). Thus (5.27) is not valid for $\ell \geq 5$.

It remains to consider the case $\ell = 3$.

Step 6. The case $\ell = 3$. We take the set S_2 constructed in Step 4. We shall show that there exists a subset in which gcd of any two elements is large. Let $0 < \theta < 1$ and

$$S_6 = \{A_i \in S_2 : A_i \geq k^{1-\theta}\}.$$

Then as seen in Step 5,

$$|S_6| \geq \epsilon k/40.$$

Denote by $b_1, \cdots, b_s$ all integers between $k^{1-\theta}$ and $ke^{(1-\epsilon/4)\log\log k}$ such that every proper divisor of b_i is $\leq k^{1-\theta}$. Observe that if $b_i \geq k^{1-\theta/2}$, every prime divisor of b_i is $\geq k^{\theta/2}$. Hence by Brun's sieve, (see Lemma 1.4.2),

$$s \ll k^{1-\theta/2} + \frac{2ke^{(1-\epsilon/4)\log\log k}}{\theta \log k}. \tag{5.29}$$

Every element of S_6 is divisible by at least one b_i. Let

$$S_7 = \{A_i \in S_6 : \exists\, b_i \text{ dividing } A_i \text{ and no other } A_j \text{ in } S_6\}.$$

Then

$$|S_7| \leq s.$$

Take

$$S_8 = S_6 \backslash S_7.$$

By (5.29) we get

$$|S_8| \geq \epsilon k/80.$$

Further, for any $A_i, A_j \in S_8$ we have

$$\gcd(A_i, A_j) \geq b_i$$

for some b_i, thus giving

$$\gcd(A_i, A_j) \geq k^{1-\theta}.$$

S_8 is the required set.

Step 7. Take S_8 with $\theta < 1/3$. For any pair (A_i, A_j) with $A_i, A_j \in S_8$, form the cubic equation

$$A_iX_i^3 - A_jX_j^3 = (i-j)d.$$

This reduces to

$$LX^3 - MY^3 = Nd \tag{5.30}$$

where

$$L = \frac{A_i}{\gcd(A_i, A_j)}, M = \frac{A_j}{\gcd(A_i, A_j)}, N = \frac{i-j}{\gcd(A_i, A_j)}$$

and L, M, N are co-prime and bounded by $k^{\theta}e^{(1-\epsilon/4)\log\log k}$ and (5.30) has a solution (X_i, X_j). The number of distinct pairs (A_i, A_j) is at least $(\epsilon k/80)^2/2$ and the number of distinct triples (L, M, N) is at most $k^{3\theta}e^{3(1-\epsilon/4)\log\log k}$. Hence there are at least

$$K = \frac{1}{2}(\epsilon/80)^2 k^{2-3\theta}e^{-3(1-\epsilon/4)\log\log k}$$

pairs (A_i, A_j) which give rise to the same cubic equation as in (5.30). Further this cubic equation has at least K distinct solutions (X_i, X_j). By Step 1, Lemma 1.2.5 and our supposition we get

$$K \le 3^{\omega(d)} < (1-\epsilon)k\frac{\log\log k}{\log k}$$

which is not possible for $k \ge k_0(\epsilon)$. This completes the case $\ell = 3$ and the proof of Theorem 5.1.3. □

5.5 Comments

(1) Theorem 5.1.1 is also known for the cases $k = 2, 3$. Győry [11] derived this as a consequence of results from Fermat type equations. We will elaborate on this aspect in later chapters.

(2) Erdős [1] observed that his method yields the following result. Let $\epsilon > 0$. Then (5.8) with $|S| \ge k - (1-\epsilon)k\frac{\log\log k}{\log k}$ implies that k is bounded by a number depending only on ϵ. Theorem 5.3.1 is a considerable improvement by Shorey. His result was further improved by Nesterenko and Shorey [12] for $\ell \ge 7$ as $\nu_\ell > 3/\ell$. For a more precise result, see their paper. Thus for $\ell \ge 7$, one may omit more than half the terms from $\{n, n+1, \ldots, n+k-1\}$ and yet, the product of the remaining terms is not a perfect power. In the same paper, they have also improved the condition on $|S|$ in Proposition 5.3.2 as $|S| \ge k\ell^{-1+\epsilon} + \pi(k) + 2$. Besides the techniques involved in Theorem 5.3.1, their improvement depends on relaxing the assumption on λ in the result of Baker in Theorem 1.2.4 even at the cost of making κ less precise. For this they used Baker's result on *simultaneous* rational approximations to certain algebraic numbers [13].

(3) We have seen that equations of the type (5.2) lead to every term in the product $\Delta(n, d, k)$ being an *almost* perfect power. It is an old question how long arithmetical progressions of equal powers of positive integers can be. In other words, with $n, d, k,$ and ℓ as in the chapter, we are asking how large k can be if

$$\text{each of } n, n+d, \ldots, n+(k-1)d \text{ is an } \ell\text{-th perfect power.} \tag{5.31}$$

There are infinitely many triples of coprime squares in arithmetical progression, but Fermat proved that there are no four squares in arithmetical progression. (If you haven't done Lemma 4.2.2 already, refer to Mordell [14, pp. 21–22]). Dénes [15] conjectured that for $\ell \geq 3$ there are no three ℓ- th powers in arithmetical progression, and proved this assertion for $3 \leq \ell \leq 30$ and sixty other prime values of ℓ. Faltings' celebrated result [16] (see Chap. 7 also) implies that for any $\ell \geq 5$ there are only finitely many triples of coprime ℓth perfect powers in arithmetical progression. By these results, it follows that if (5.31) holds then k is bounded by some number depending only on ℓ. Shorey and Tijdeman [17, Theorem 1] showed that (5.31) implies that k is bounded by a number depending only on $\omega(D_1)$. Recall that D_1 is the maximal divisor of d whose prime factors are $\equiv 1 \pmod{\ell}$.

(4) In this chapter we investigated arithmetic progressions consisting of almost perfect powers in the 'homogeneous' case as Lajos Hajdu terms it [18], i.e., one is interested in arithmetic progressions of the shape

$$a_0 x_0^{\ell}, \ldots, a_{k-1} x_{k-1}^{\ell} \text{ with } a_i, x_i \in \mathbb{Z}, 0 \leq i < k$$

with some fixed integer $\ell \geq 2$, such that the coefficients a_i are'restricted' in some sense. The 'inhomogeneous' case was studied by Hajdu [18] and later by Hajdu and Tengly [19]. For instance, suppose

$$x_1^{\ell_1}, \ldots, x_t^{\ell_t}$$

is an arithmetic progression in $\mathbb{Z}$ with $\gcd(x_1^{\ell_1}, x_2^{\ell_2}) = 1$ and $2 \leq \ell_i \leq L, 1 \leq i \leq t$, then t is bounded by some number depending only on L. Further, if all $\ell_i \in \{2, p\}$ where p is a prime, then $t \leq 6$. On the other hand, it is known that there exist exponents $\ell_1, \ell_2, \ell_3 \geq 2$ for which there are infinitely many arithmetic progressions as above and in this case, the exponents satisfy

$$\frac{1}{\ell_1} + \frac{1}{\ell_2} + \frac{1}{\ell_3} \geq 1.$$

We will address this problem in later chapters as well.

Exercise 5

(1) Let $d > 1$ be any integer. Find out in how many ways the ideal $[d]$ can be written as product of n pairwise coprime ideals in $\mathbb{Q}(e^{2\pi i/3})$.

References

1. P. Erdős, On the product of consecutive integers III. Indag. Math. **17**, 85–90 (1955). [**1, 4, 5**]
2. P. Erdős, On a Diophantine equation. J. Lond. Math. Soc. **26**, 176–178 (1951). [**4, 5**]
3. P. Erdős, J.L. Selfridge, The product of consecutive integers is never a power. Illinois J. Math. **19**, 292–301 (1975) [**2, 5**]
4. R. Marszalek, On the product of consecutive elements of an arithmetic progression. Monatsch. Math. **100**, 215–222 (1985). [**5**]
5. T.N. Shorey, R. Tijdeman, Perfect powers in products of terms in an arithmetical progression. Compositio Math. **75**, 307–344 (1990). [**4, 5, 6**]
6. N. Saradha, On perfect powers in products with terms from arithmetic progressions. Acta Arith. **82**, 147–172 (1997). [**2–5**]
7. T.N. Shorey, R. Tijdeman, Perfect powers in products of terms in an arithmetical progression (III). Acta Arith. **61**, 391–398 (1992). [**5**]
8. P. Das, S. Laishram, N. Saradha, Cubes in products of terms from an arithmetic progression. Acta Arith. **184**, 117–126 (2018). [**5**]
9. T.N. Shorey, Perfect powers in products of integers from a block of consecutive integers. Acta Arith. **49**, 71–79 (1987). [**4**]
10. T.N. Shorey, Perfect powers in values of certain polynomials at integer points. Math. Proc. Cambridge Philos. Soc. **99**, 268–273 (1986). [**1**]
11. K. Győry, On the diophantine equation $n(n+1)\cdots(n+k-1) = bx^\ell$. Acta Arith. **83**, 87–92 (1998). [**5, 7**]
12. Y.V. Nesterenko, T.N. Shorey, Perfect powers in products of integers from a block of consecutive integers (II). Acta Arith. **76**, 191–198 (1996). [**5**]
13. A. Baker, Simultaneous rational approximations to certain algebraic numbers. Proc. Cambridge Philos. Soc. **63**, 693–702 (1967). [**5**]
14. L. J. Mordell, *Diophantine Equations*, vol. 30 (Academic Press, 1969). [**5**]
15. P. Dénes, *Über die Diophantische Gleichung* $x^\ell + y^\ell = cz^\ell$. Acta Math. **88**, 241–251 (1952). [**1, 5**]
16. G. Faltings, Endlichkeitssätze für abelsche Varieäten über Zahlkörpern. Invent. Math. **71**, 349–366 (1983). [**5**]
17. T.N. Shorey, R. Tijdeman, Perfect powers in arithmetical progression (II). Compositio Math. **82**, 107–117 (1992). [**5**]
18. L. Hajdu, Perfect powers in arithmetic progression. A note on the inhomogeneous case. Acta Arith. **113**, 343–349 (2004). [**5**]
19. L. Hajdu, S. Tengly, Arithmetic progressions of squares, cubes and n-th powers. Functiones et Approx. **41**(2), 129–138 (2009). [**5**]

Chapter 6
No Perfect Powers for Infinitely Many d

$$d \neq 2^{\alpha}3^{\beta}5^{\gamma} > 1$$

Let us recall the equation (5.2) and some notation from Sect. 2.4. Let

$$\Delta(n, d, k) = by^{\ell} \text{ in integers } n > 0, d > 1, k > 1, y > 1, b \geq 1, \ell \geq 3, \ell\text{-prime}.$$

For any integer $n > 1$, we write

$$n = n_1 n_2$$

where n_1 is the maximal divisor of n such that every prime divisor of n_1 is congruent to 1 $\pmod{\ell}$. Consequently, every prime divisor of n_2 is incongruent to 1 $\pmod{\ell}$ and $\gcd(n_1, n_2) = 1$. We shall call n_1 as the ℓ—part of n. When $n = d$ we write

$$d = D_1 D_2$$

where D_1 is the ℓ—part of d and every prime divisor of D_2 is incongruent to 1 $\pmod{\ell}$. When $d = 1$ we take $D_1 = D_2 = 1$.

$$\theta = \begin{cases} 1 \text{ if } \ell \nmid d \\ 1/\ell \text{ if } \ell | d. \end{cases}$$

We take $\ell' = 2$ in Lemma 2.4.4. For any $\kappa > 0$, let

$$\kappa_0 = \min\left(\frac{\ell}{2(\kappa+1)^{(\ell-2)/\ell}}, \frac{\kappa}{(\kappa+1)2^{1/\ell}}\right).$$

We have seen in Corollary 5.2.1 that

S. Natarajan, *Perfect Powers—An Ode to Erdős*, Infosys Science Foundation Series,
https://doi.org/10.1007/978-981-96-2599-4_6

suppose (5.2) holds, then

$$D_1 > \kappa_0 \frac{\theta}{k}(n + k - 1)d)^{(\ell-2)/\ell}. \tag{6.1}$$

We know that $(n + (k - 1)d \geq (k + 1)^\ell$. So we get

$$D_1 > \kappa_0 \theta k^{\ell-3}.$$

Here the sticky point is κ_0. It may turn out to be very very small so that the right-hand side is < 1. This is of no use since the least value of D_1 is 1. Yet another weak point is $\ell = 3$. So it is desirable to make κ_0 more explicit and find a bound which includes the case $\ell = 3$. Shorey [1, Theorem 2(i)] showed that indeed

$$D_1 > 1 \text{ if } k \text{ is sufficiently large.}$$

Shorey and Tijdeman [2, Theorem 2] showed that

$$D_1 \gg k^{\ell-2}$$

but the constant involved in the $\gg$ sign turns out to be too small that the above bound is useless for small values of k. What is the consequence of such results? Suppose (6.1) implies $D_1 > 1$. It means d is divisible by a prime $\equiv 1 \pmod{\ell}$. Since $\ell \geq 3$, we get $d \geq 7$ so that we conclude that

(5.2) with $d = 2^\alpha 3^\beta 5^\gamma > 1$ *has no solution for k large.*

Thus there are infinitely many values of d for which (5.2) does not hold. As another use we see that as k and ℓ increase, the right-hand side of the inequality increases for a reasonable κ_0. This then gives a large lower bound for d since $d \geq D_1$. This involves improving the lower bound for $n + (k - 1)d$. Also it is desirable to have results which are true for small values of k too. Saradha and Shorey [3, Theorem 2] proved such a result which will be presented in this chapter (Theorems 6.1.1 and 6.1.2). The proofs are highly computational.

6.1 Statement of the Results

In this chapter, we prove the following results.

Theorem 6.1.1 *Equation (5.2) implies that*

$$D_1 > 1 \textit{ for } k \geq 4.$$

Table 6.1 Triples (ℓ, γ, δ) of Theorem 6.1.2

ℓ	γ	δ	δ
		$k \in E$	$k \notin E$
3	0.42	1/3	1/3
5	0.62	1/5 for $k = 4$ *or* $5 \nmid d$	7/5
	0.71	7/5 for $k \neq 4$ & $5 \mid d$	
7	0.73	9/7	17/7
11	0.85	25/11	25/11
13	1.11	43/13	57/13
17	1.63	91/17	145/17
19	2.32	141/19	201/19
23	3.56	217/23	337/23
29	6.83	361/29	541/29
≥31	0.3	$\ell - \lfloor 2\ell^{2/3} \rfloor + 1$	$\ell - \lfloor 2.3\ell^{1/2} \rfloor - 1$ if $k < 30$
	9.98		$\ell - \lceil 7\ell^{1/4} \rceil - 1$ if $k \geq 30$

Thus we now have the better result that

(5.2) with $d = 2^\alpha 3^\beta 5^\gamma > 1$ *has no solution for* $k \geq 4$.

We shall also prove a weak but explicit version of the result of Shorey and Tijdeman described in the introduction.

Theorem 6.1.2 *Let* $E = \{4, 5, 6, 7, 8, 24, 29\}$. *Equation (5.2) with* $k \geq 4$ *implies that*

$$D_1 > \gamma\theta k^\delta$$

where (ℓ, γ, δ) *are given in Table 6.1.*

The bound for D_1 when $\ell \geq 11$ is better than the one given in [3, Theorem 3]. The above theorem is valid for all $k \geq 4$. When k is large, we get better result as follows.

Theorem 6.1.3 *Equation (5.2) implies that*

$$D_1 > \theta k^{\ell - 8/3} \; if \; k \geq 7174.$$

This chapter depends on many computations which can be carried out easily using MATHEMATICA or any other package available. The method used is capable of giving better results for specific k and ℓ. In fact, we have

$$D_1 > \beta_0 \theta k^{\ell - \ell' - \ell'/\ell}$$

with β_0 as given in Table 6.3 and ℓ' as in Table 6.2.

Table 6.2 Choices of (ℓ, ℓ', κ) for computing κ_0

ℓ	$k \in E$	$k \notin E$
–	ℓ'; κ	ℓ'; κ
3	2; 8	2; 8
5	4; 13 if $k = 4$ or $5 \nmid d$	3; 7
	3; 7 if $k \neq 4$ and $5 \mid d$	
7	5; 8	4; 7
11	8; 6	8; 6
13	9; 6	8; 6
17	11; 5	8; 6
19	11; 5	8; 6
23	13; 4	8; 6
29	16; 7	10; 6 if $k < 30$
		9; 6 if $k \geq 30$.
≥ 31	$\lfloor 2\ell^{2/3} \rfloor - 2$; 7	$\lfloor (2.3)\ell^{1/2} \rfloor$; 7 if $k < 30$
		$\lceil 7\ell^{1/4} \rceil$; 6 if $k \geq 30$.

Table 6.3 Value of β_0 in the lower bound of D_1

ℓ	$k < 30$	$k < 30$	$k \geq 30$	$k \geq 30$
–	β	$\beta_0 = (0.7)\beta^{(\ell-\ell')/\ell}$	β	$\beta_0 = (0.7)\beta^{(\ell-\ell')/\ell}$
3	0.2646	0.4494	0.2164	0.4202
5	0.7629	0.6281	1.0629	0.7172
7	1.1920	0.7360	1.2366	0.7666
11	2.9103	0.8500	5.3025	1.1032
13	4.5474	1.1155	6.1452	1.4072
17	11.1072	1.6370	6.1695	1.8342
19	17.3472	2.3274	15.9947	3.4844
23	42.3516	3.5680	83.7572	12.5673
29	161.559	6.8397	647.677	60.8087
≥ 31	0.1666, $k \in E$	0.3115	359.57	9.9844
	0.4117, $k \notin E$	0.4063		

6.2 Computation of κ_0

Since $d > 1$ and $k \geq 4$, we know that

$$n + (k-1)d \geq p^{\ell}_{\pi(k)+1} \geq (k+1)^{\ell}. \tag{6.2}$$

Further by Lemma 2.4.1, we find that

$$|T(A)| = |T(a)| = k,$$

i.e., all the A_i's are distinct and so are the a_i's provided

$$D_1 \leq (0.84)\theta k^{\ell-2} \tag{6.3}$$

by the modified version in Exercise 2 (8). This is our starting point. By imposing stricter condition on D_1, it is possible to bound k. Already we have seen this idea in Stage 1 of Theorem 5.1.2 by taking $D_1 = 1$ and $\ell' = 2$. This argument in fact gives us $k \leq 7268$ provided

$$D_1 \leq \theta k^{\ell-8/3}.$$

This is Lemma 6.4.1. The main strategy is to make full use of Lemma 2.4.4 and then contradict the inequalities (i) and (ii) of Lemma 2.2.1. There are three main parameters governing the condition on D_1, viz., k, ℓ and ℓ'. We divide k and ℓ into four parts, viz.,

$$\begin{cases} \text{(i) } k < 30, \ell < 30 & \text{(ii) } k < 30, \ell \geq 30 \\ \text{(iii) } k \geq 30, \ell \geq 17 & \text{(iv) } k \geq 30, 3 \leq \ell \leq 13. \end{cases} \tag{6.4}$$

The above division enables us to choose ℓ' more economically. First let us look at

$$\kappa_0 = \min\left(\frac{\ell}{\ell'(\kappa+1)^{(\ell-\ell')/\ell}}, \frac{\kappa}{(\kappa+1)\ell'^{1/\ell}}\right)$$

where κ is a positive real number. Note that κ_0 does not depend on k. For $\ell < 30$ we give specific values of ℓ' and for $\ell \geq 30$ we specify ℓ' as a function of ℓ. By using the values of κ given in Table 6.2, one can check that

$$\kappa_0 > (0.7).$$

We have one more instance to compute κ_0 when $\ell' = 2$ and $\ell \geq 3$. First we take $\ell = 3$ and $\kappa = 8$ we have already seen that $\kappa_0 > (0.7)$. Let $\ell \geq 5$. It is easy to see that κ_0 increases as ℓ increases. So we compute κ_0 at $\ell = 5$ with $\kappa = 6$ to find $\kappa_0 > (0.7)$. Thus

$$\kappa_0 > (0.7) \text{ for all } \ell \geq 3 \text{ when } \ell' = 2. \tag{6.5}$$

6.3 Improving the Bound for $n + (k - 1)d$

The next task is to improve (6.2). Let us write

$$n + (k - 1)d \geq p^{\ell}_{\pi(k)+1} := \alpha_0 k^{\ell+1}.$$

The first lemma is a formalization of the techniques already we have encountered in Lemma 4.4.1 and Theorem 5.3.1.

Lemma 6.3.1 *Suppose Eq. (5.2) with $k \geq 4$ and (6.3) hold. Then*

(i) $n + (k - 1)d \geq \alpha_1 k^{\ell+1}$ *where*

$$\alpha_1 = 1 - \frac{\log(k-1)!}{k(\ell \log p_{\pi(k)+1} - \log k)} - \frac{\pi(k)+1}{k}.$$

(ii) $n + (k - 1)d \geq \alpha_2 k^{\ell+1}$ *where*

$$\alpha_2 = 1 - \frac{\log k}{\ell \log k + \log \alpha_1'} - \frac{\pi(k)+1}{k} \text{ with } \alpha_1' = \max(\alpha_0, \alpha_1).$$

(iii) $n + (k - 1)d \geq p^{\ell}_{\pi(k)+\lceil \alpha_2' k \rceil} = \alpha_3 k^{\ell+1}$ *where*

$$\alpha_3 = p^{\ell}_{\pi(k)+\lceil \alpha_2' k \rceil} / k^{\ell+1} \text{ with } \alpha_2' = \max(\alpha_1, \alpha_2).$$

(iv) $n + (k - 1)d \geq \alpha_4 k^{\ell+1}$ *where*

$$\alpha_4 = 1 - \frac{1}{\ell - 1} - \frac{\pi(k)+1}{k}.$$

(v) $n + (k - 1)d \geq \alpha_5 k^{\ell+1}$ *where*

$$\alpha_5 = \max\left\{\alpha_4, 1 - \frac{\log k}{\ell \log k + \log \alpha_4} - \frac{\pi(k)+1}{k}\right\}.$$

(vi) $n + (k - 1)d \geq p^{\ell}_{\pi(k)+\lceil \alpha_5 k \rceil} := \alpha_6 k^{\ell+1}$ *where*

$$\alpha_6 = p^{\ell}_{\pi(k)+\lceil \alpha_5 k \rceil} / k^{\ell+1}.$$

(vii) $n + (k - 1)d \geq \alpha_7 k^{\ell+1}$ *where*

$$\alpha_7 = (\alpha_5 - \delta)\,(\delta \log(\delta k))^{\ell} \text{ and } 0 \leq \delta < \alpha_5.$$

Note: (*i*)–(*iii*) are useful for specific values of k and when k is small. (*iv*) − −(*vii*) are useful for large values of k.

Proof As usual we take $S = \{n, n+d, \cdots, n+(k-1)d\}$. By (6.3), we have $|T(A)| = k$. Let

$$S_1 = \{A_i \in T(A) : X_i \neq 1\}$$

and

$$S_2 = T(A) \backslash S_1.$$

For any $A_i \in S_2$, $X_i = 1$ and hence $A_i = n + id$ and so for $i \neq 0$, $A_i > \frac{p^{\ell}_{\pi(k)+1}}{k} > k^{\ell-1}$. By Erdős argument, there exists a set $S_3 \subset S_2$ with $|S_3| \geq |S_2| - \pi(k)$ such that

$$\prod_{A_i \in S_3} A_i \leq (k-1)! \tag{6.6}$$

Thus

$$k^{(\ell-1)(|S_2|-\pi(k)-1)} < \left(\frac{p^{\ell}_{\pi(k)+1}}{k}\right)^{|S_2|-\pi(k)-1} \leq (k-1)! < k^k$$

giving

$$|S_2| < \frac{\log(k-1)!}{\ell \log p_{\pi(k)+1} - \log k} + \pi(k) + 1$$

and more simply

$$|S_2| < \frac{k}{\ell - 1} + \pi(k) + 1.$$

Hence

$$|S_1| > k - \frac{\log(k-1)!}{\ell \log p_{\pi(k)+1} - \log k} - \pi(k) - 1 = \alpha_1 k$$

and

$$|S_1| > k - \frac{k}{\ell - 1} - \pi(k) - 1 = \alpha_4 k.$$

Since for all the A_i's in S_1, $X_i > k$ we get

$$n + (k-1)d \geq \alpha_1 k^{\ell+1} \text{ or } \alpha_4 k^{\ell+1}.$$

This proves (*i*) and (*iv*).

Let us have the same notation as in (*i*) or (*iv*). Let $\alpha' = \max(\alpha_0, \alpha_1)$ or α_4. Then $n + (k-1)d \geq \alpha' k^{\ell+1}$. Hence for $A_i \in S_2$, $i \neq 0$, we have $A_i > \alpha' k^{\ell}$ and this yields

$$|S_2| < \frac{k \log k}{\ell \log k + \log \alpha'} + \pi(k) + 1$$

giving

$$|S_1| > k \left(1 - \frac{\log k}{\ell \log k + \log \alpha'} - \frac{\pi(k) + 1}{k}\right).$$

Hence

$$|S_1| \geq \lceil \alpha_2 k \rceil \text{ or } \lceil \alpha_5 k \rceil$$

giving

$$n + (k-1)d \geq \alpha_2 k^{\ell+1} \text{ or } \alpha_5 k^{\ell+1},$$

thus proving (*ii*) and (*v*).

We have $|S_1| \geq \lceil \alpha_2' k \rceil$ where $\alpha_2' = \max(\alpha_1, \alpha_2)$. Consider the X_i's corresponding to $A_i \in S_1$ and arrange them in the increasing order as

$$X_1 < X_2 < \cdots < X_{\lceil \alpha_2' k \rceil} < \cdots .$$

Since each X_i is divisible by distinct primes $> k$ we derive that

$$X_{\lceil \alpha_2' k \rceil} \geq p_{\pi(k) + \lceil \alpha_2' k \rceil}$$

proving (*iii*). Proof of (*vi*) is similar.

Consider the sequence of X_i's as above. Let us omit the first $\delta k - \pi(k) < \alpha_5 k$ of these. Thus we are left with

$$X_{\lceil \delta k - \pi(k) \rceil} < \cdots < X_{\lceil \alpha_5 k \rceil} < \cdots .$$

Note that each X_i is $> p_{\pi(k) + \lceil \delta k - \pi(k) \rceil} > p_{\delta k}$. Arrange the A_i's corresponding to these X_i's in the increasing order. There are at least $\left(\alpha_5 - \delta + \frac{\pi(k)}{k}\right) k$ of them. We conclude that

$$\begin{aligned} n + (k-1)d &\geq \left(\alpha_5 - \delta + \frac{\pi(k)}{k}\right) k (X_{\lceil \delta k - \pi(k) \rceil})^{\ell} \\ &\geq \left(\alpha_5 - \delta + \frac{\pi(k)}{k}\right) k (p_{\delta k})^{\ell} \\ &\geq \left(\alpha_5 - \delta + \frac{\pi(k)}{k}\right) k \big(\delta k \log(\delta k)\big)^{\ell} \\ &> (\alpha_5 - \delta)\, (\delta \log(\delta k))^{\ell}\, k^{\ell+1} = \alpha_7 k^{\ell+1}. \end{aligned}$$

This proves (*vii*).

Remark. Suppose there exists a prime $p < k$ dividing d. Then inequality in (6.6) can be replaced with

$$\prod_{A_i \in S_3} A_i \leq (k-1)! p^{-\nu_p((k-1)!)} \leq (k-1)! p^{-k/(p-1)}.$$

Using this, the values of α_i's in Lemma 6.3.1 can be improved.

Our aim is to apply the above lemma to find

$$n + (k-1)d \geq \beta k^{\ell+1}$$

in the four cases in (6.4). We shall explain by means of examples all the four cases. Table 6.3 gives the value of β in all the cases.

Case (i) $k < 30, \ell < 30$. Let $\ell = 3$. Consider $k = 4$. By Lemma 6.3.1(i)–(iii), we get

$$\alpha_0 \geq (0.4882), \alpha_1 \geq (0.1198), \alpha_2 \geq -(0.1528) \text{ and } \alpha_2' \geq (0.1198)$$

and so $\alpha_3 \geq (0.4882)$. Thus $n + (k-1)d \geq (0.4882) \times 4^4$. By such computation, we see that for all $k < 30$, $n + (k-1)d \geq (0.2646)k^4$. The minimum value of β occurs for $k = 6$. Similarly for every $\ell < 30$, we compute the minimum value of β for $4 \leq k < 30$. The minimum occurs at $k = 4$ for $5 \leq \ell < 30$.

Case (ii) $k < 30, \ell \geq 30$. Let $k \in E$. By (iv) and (v) of Lemma 6.3.1 we get $\alpha_5 \geq (0.1666)$ by taking $\ell = 31$ and exact values of $\pi(k)$. The minimum value occurs at $k = 5$. As α_5 is an increasing function of ℓ, this value holds good for all $\ell > 31$ and so $n + (k-1)d \geq (0.1666)k^{\ell+1}$, $k \in E$. Similarly we compute α_5 for all $k < 30$, $k \notin E$ and find that $n + (k-1)d \geq (0.4117)k^{\ell+1}$. The minimum value of β occurs at $k = 9$.

Case (iii) $k \geq 30, \ell \geq 17$. Let $17 \leq \ell \leq 29$. Here we compute α_4 and α_5 with exact values of $\pi(k)$ for $k = 30, 31, 32$ and then compute α_7 with $\delta = (0.5)$. For $k \geq 33$ we compute α_4, α_5 and α_7 with approximate values of $\pi(k)$. As an example, let us take $\ell = 17$. With exact value of $\pi(k)$ for $k = 30, 31, 32$, we find $\alpha_4 \geq (0.5504)$ and $\alpha_5 \geq (0.5534)$, and $\alpha_7 \geq 11.3406$ with $\delta = (0.5)$. For all $k \geq 33$ we use approximate value for $\pi(k)$. Thus $\alpha_4 \geq (0.5168)$, $\alpha_5 \geq (0.5198)$ and lastly $\alpha_7 \geq 6.1695$ with $\delta = (0.5)$. So we conclude that

$$\beta \geq 6.1695 \text{ if } \ell = 17.$$

The other values of $\ell \leq 29$ are similar. Let $\ell \geq 31$. For $k \geq 30$ with approximate value of $\pi(k)$ we get

$$\alpha_4 \geq (0.5289), \alpha_5 \geq (0.5298), \alpha_7 \geq 359.57, \delta = (0.5).$$

Thus

$$\beta \geq 359.57 \text{ for } \ell \geq 31.$$

Case (*iv*) $k \geq 30, 3 \leq \ell \leq 13$. Let

$$k_0 = \begin{cases} 60 \text{ if } \ell = 3 \\ 130 \text{ if } \ell = 5 \\ 65 \text{ if } \ell = 7 \\ 46 \text{ if } \ell = 11 \\ 40 \text{ if } \ell = 13. \end{cases}$$

If $\ell = 3$, we get $\alpha_5 \geq (0.2718)$ for $k \geq 60$. When $30 \leq k < 60$ we compute $\alpha_6 \geq (0.2164)$. Hence

$$\beta \geq (0.2164) \text{ for } \ell = 3 \text{ and } k \geq 30.$$

Let $5 \leq \ell \leq 13$. Then for $k \geq k_0$ we find α_7 with approximate value of $\pi(k)$ and $\delta = (0.5)$. For $k < k_0$ we compute α_6 with exact value of $\pi(k)$. For example, let $\ell = 13$. For $k \geq 40$, we get $\alpha_4 \geq (0.5267), \alpha_5 \geq (0.5321)$ and $\alpha_7 \geq 6.1452$. For $30 \leq k < 40$, with exact value of $\pi(k)$ we get $\alpha_6 \geq 259684$. Hence

$$\beta \geq 6.1452 \text{ for } \ell = 13 \text{ and } k \geq 30.$$

All the above calculations are recorded in Table 6.3.

We have occasion to find α_5 when $k \geq 7174, \ell \geq 3$. Then $\alpha_4 \geq (0.3710)$ and $\alpha_5 \geq (0.5248)$. We compute α_7 with $\delta = (0.4)$ to get

$$\alpha_7 \geq 4.03. \tag{6.7}$$

We hope the reader understands the power of Lemma 6.3.1. While forming Table 6.3, we want to be as uniform as possible with respect to k and ℓ. For given k and ℓ, the lemma can yield much better result. To illustrate, let us take $\ell = 11, k = 30$. From Table 6.3 we get

$$n + (k-1)d \geq (5.3025)30^{12}.$$

On the other hand, we compute $\alpha_6 \geq 26046.8$ so that

$$n + (k-1)d \geq (26046.8)30^{12}.$$

6.4 Some Lemmas to Facilitate the Proof of the Theorems

From the previous two sections, we get

$$\kappa_0 \frac{\theta}{k}(n+(k-1)d)^{(\ell-\ell')/\ell} \geq (0.7)\theta\beta^{(\ell-\ell')/\ell} k^{\ell-\ell'-\ell'/\ell}.$$

Thus if

$$D_1 \leq \beta_0 \theta k^{\ell-\ell'-\ell'/\ell}, \tag{6.8}$$

with β_0 as in Table 6.3, then the assertions of Lemmas 2.4.4 and 2.2.1 are valid. So we need only to show that these assertions lead to a contradiction. This is the strategy that will be followed.

Lemma 6.4.1 *Suppose Eq. (5.2) holds with*

$$D_1 \leq \theta k^{\ell-8/3}.$$

Then $k \leq 7173$.

Proof Let $k \geq 7174$. By our supposition on D_1, Eq. (6.3) is satisfied and hence all A_i's are distinct. So we may apply Lemma 6.3.1. By (6.7), we have

$$n+(k-1)d \geq (4.03)k^{\ell+1}.$$

We know by (6.5) and (6.1) that $\kappa_0 > (0.7)$ and

$$\begin{aligned} D_1 > \kappa_0 \frac{\theta}{k}(n+(k-1)d)^{(\ell-2)/\ell} &> \frac{\theta}{k}(0.7)(4.03)^{1/3} k^{(\ell+1)(\ell-2)/\ell} \\ &> (1.11)\theta k^{\ell-2-2/\ell} \end{aligned}$$

which contradicts our assumption on D_1. So we conclude that $k \leq 7173$. □

Note.

(i) The above lemma says that if Eq. (5.2) holds with $k \geq 7174$, then

$$D_1 > \theta k^{\ell-8/3} > 1.$$

So Theorem 6.1.1 is true for $k \geq 7174$. We need to cover small values of k.

(ii) Let $\ell \leq 101$. Then we can replace the condition on D_1 in the above lemma by

$$D_1 \leq \theta k^{\ell-2-2/\ell}.$$

(Find out why we need the condition $\ell \leq 101$.) Thus for $k \geq 7174$ and $\ell \leq 101$, we have

$$D_1 > \theta k^{\ell-2-2/\ell} > 1.$$

For instance, when $k = 7174$ and $\ell = 19$, by the lemma, we get $D_1 > \theta(9.49) \times 10^{62}$ while the above bound gives $D_1 > \theta(1.38) \times 10^{65}$.

Lemma 6.4.2 *Suppose (5.2) holds with $\ell \geq 3$ and $k < 30$. Then*

$$D_1 > \beta_0 \theta k^{\ell-\ell'-\ell'/\ell}$$

where ℓ, ℓ' are as given in Table 6.2 and β_0 as in Table 6.3.

Proof Let $k < 30$ and $\ell \geq 3$. Assume that

$$D_1 \leq \beta_0 \theta k^{\ell-\ell'-\ell'/\ell}.$$

As noted in (6.8), the assertions of Lemmas 2.4.4 and 2.2.1 are valid. Let $\ell = 3$. Then property P_2 holds. We find that $f_0(k, 2) \geq 4$ for $4 \leq k < 30$. But then the inequality in Lemma 2.2.1 (i) does not hold. Thus we conclude that for $\ell = 3$ the assertion of the lemma holds. Similarly we check all the ℓ values ≤ 29 using ℓ' given by Table 6.2. Note here that when $\ell = 5, 5 \mid d$ and $k \in E\backslash\{4\}$, then 5 does not divide any of the a_i's and hence $f(k, 2) \geq 5$. So we may take $\ell' = 3$ to show that the inequality in Lemma 2.2.1 (i) does not hold. This is also true when $k \notin E$. Now let $\ell \geq 31$. Then

$$\ell' = \lfloor 2\ell^{2/3} \rfloor - 2 \text{ if } k \in E;\ \ell' = \lfloor (2.3)\ell^{1/2} \rfloor \text{ if } k \notin E.$$

Since $f_0(k, 2) \geq 4$ for $k \in E$ we get

$$\begin{aligned}\binom{\lfloor 2\ell^{2/3} \rfloor + 1}{3} &\geq (2\ell^{2/3})(2\ell^{2/3} - 1)(2\ell^{2/3} - 2)/6 \\ &\geq \ell^2 + \frac{\ell^2 - 6\ell^{4/3}}{3} \\ &> \ell^2\end{aligned}$$

as $\ell \geq 31$. For $k \notin E$, we have $f_0(k, 2) \geq 5$ and so

$$\begin{aligned}\binom{\lfloor 2.3\ell^{1/2} \rfloor + 4}{4} &\geq (2.3\ell^{1/2} + 3)(2.3\ell^{1/2} + 2)(2.3\ell^{1/2} + 1)(2.3\ell^{1/2})/24 \\ &\geq 27.98\ell^2/24 \\ &> \ell^2.\end{aligned}$$

Thus the inequality in Lemma 2.2.1 (i) is not satisfied and hence the assertion of the lemma holds for $\ell \geq 31$. □

Next we deal with $k \geq 30$. In the following lemma we bring down the value of k from 7174 by weakening the condition on D_1. We now take $\ell' \geq 8$. This is possible only for $\ell \geq 11$.

Lemma 6.4.3 *Suppose (5.2) holds with $k \geq 30$, $\ell \geq 11$ and*

$$D_1 \leq \beta_0 \theta k^{\ell-\ell'-\ell'/\ell} \tag{6.9}$$

where $\ell' \geq 8$, then $k \leq 1352$.

The proof of this lemma is similar to Theorem 5.1.2-Stage 1. We shall use the same values of m and T as in Table 5.1. Thus the values of $|V|$ and μ remain the same. Since $\ell' \geq 8$, $|U^{(0)}|$ changes. Also observe that it is enough to compute the lower bound for a_h with $\ell' = 8$. For instance, let U be the set of all integers ≤ 16 composed of the primes 2 and 3. Thus $U = \{1, 2, 3, 4, 6, 8, 9, 12, 16\}$. Then $U^{(0)} = \{(1, 1), (2, 1), (3, 1)\}$. This will alter the values of ν and Least h. Table 6.4 is analogous to Table 5.1 with these changed values.

Proof First, note that by (6.9), D_1 satisfies the condition in Lemma 6.4.1. Hence we get $k < 7174$. Let $k > 1352$. By the given condition on D_1, the assertion of Lemma 2.4.4 is valid with $\ell' \leq 8$. Thus $T(a)$ has the property $P_{\ell'}$ with $1 \leq \ell' \leq 8$ and so the bounds for a_h given in Table 6.4 are valid. We know that all a_i's are distinct. Arrange the a_i's in the increasing order as

$$a_1' < a_2' < \cdots < a_k'.$$

By Erdős argument we see that there exists a set $S' \subset T(a)$ of cardinality $\geq k - \pi(k)$ such that

$$\prod_{a_i' \in S'} a_i' \leq (k-1)!$$

Table 6.4 Values connected to sets U and V when $\ell' \geq 8$

No.	m	T	$\|U^{(0)}\|$	$\|V\|$	μ	ν	Least h
1	–	–	–	–	1	0	1
2	2	16	3	0.3924X+2	2.5484	5	9
3	3	16	4	0.3424 X+3	2.9205	7	21
4	6	36	8	0.2431 X+10	4.1135	18	45
5	7	36	9	0.2356 X+16	4.2444	25	245
6	8	39	10	0.2279 X+24	4.3878	34	301
7	9	42	11	0.2224 X +39	4.4964	50	697

Using Table 6.4 and the above inequality, we see that

$$\log\left(8!(2.5484)^{12}\frac{(15)!}{3!}(2.9205)^{24}\frac{37!}{13!}(4.1135)^{200}\frac{(226)!}{26!}\right.$$
$$\left.\times(4.2444)^{56}\frac{(275)!}{219!}(4.3878)^{396}\frac{(662)!}{266!}(4.4964)^{k-\pi(k)-697}\frac{(k-\pi(k)-50)!}{646!}\right)$$
$$-\log(k-1)! \leq \log\left(\prod_{a_i'\in S'} a_i'\right) - \log(k-1)! \leq 0.$$

For $1353 \leq k < 7174$, we check using exact values of $\pi(k)$ that the above inequality is not valid. Thus $k \leq 1352$.

Lemma 6.4.4 *Suppose (5.2) holds with* $k \geq 30$ *and* $\ell \geq 11$. *Then*

$$D_1 > \beta_0\theta k^{\ell-\ell'-\ell'/\ell}$$

where ℓ' *is*

$$\ell' = \begin{cases} 8 \text{ if } 11 \leq \ell \leq 23 \\ 9 \text{ if } \ell = 29 \\ \lceil 7\ell^{1/4}\rceil \text{ if } \ell \geq 31. \end{cases}$$

Proof Suppose

$$D_1 \leq \beta_0\theta k^{\ell-\ell'-\ell'/\ell}.$$

We see that D_1 satisfies (6.9). Hence we derive that $k \leq 1352$, by Lemma 6.4.3. Let $k \geq 30$. By the supposition on D_1 the assertion of Lemma 2.4.4 is valid and

$$\binom{f_0(k,m)+\ell'-1}{\ell'} \leq \ell^m. \tag{6.10}$$

Using (2.9), we check that $f(k,m) \geq f_0(k,m)$ where

$$\begin{cases} f_0(k,4) \geq 16, 30 \leq k \leq 310;\ f_0(k,5) \geq 27, 311 \leq k \leq 611; \\ f_0(k,6) \geq 45, 612 \leq k \leq 1102;\ f_0(k,7) \geq 81, 1103 \leq k \leq 1352. \end{cases} \tag{6.11}$$

First we take $11 \leq \ell \leq 23$. Then $\ell' = 8$. Let $30 \leq k \leq 310$. Since $f_0(k,4) \geq 16$, by (6.11), we have

$$\binom{23}{8} \leq \ell^4 \leq 23^4$$

which is not valid. Thus $k \geq 311$. Next we consider $311 \leq k \leq 611$. Then $f_0(k,5) = 27$. We check that

$$\binom{34}{8} > 23^5.$$

Hence $k > 612$. Similarly we check that

$$\binom{52}{8} > 23^6 \text{ and } \binom{88}{8} > 23^7.$$

Thus $k > 1352$, a contradiction. By similar argument we rule out $\ell = 29$ with $\ell' = 9$. Next, we take $\ell \geq 31$. Let $30 \leq k \leq 310$. Then

$$\binom{\lceil 7\ell^{1/4} \rceil + 15}{16} \leq \ell^4.$$

This means

$$(\lceil 7\ell^{1/4} \rceil + 15)(\lceil 7\ell^{1/4} \rceil + 14) \cdots (\lceil 7\ell^{1/4} \rceil + 1)\lceil 7\ell^{1/4} \rceil < 16!\ell^4.$$

This implies that

$$(7\ell^{1/4} + 15)(7\ell^{1/4} + 14) \cdots (7\ell^{1/4} + 1)(7\ell^{1/4}) < 16!\ell^4$$

which gives

$$(7\ell^{1/4})^{16} < 16!\ell^4.$$

This is not true. Thus $k > 310$. Let $311 \leq k \leq 611$. Then

$$(\lceil 7\ell^{1/4} \rceil + 26) \cdots (\lceil 7\ell^{1/4} \rceil + 1)(\lceil 7\ell^{1/4} \rceil) < 27!\ell^5.$$

Thus

$$(7\ell^{1/4})^{27} < 27!\ell^5.$$

This gives $\ell \leq 960$. Now by calculating the exact value of ℓ', we check for each value of ℓ in the range $31 \leq \ell \leq 960$, ℓ prime, that the inequality (6.10) does not hold. Thus $k > 611$. By using similar argument as above, we show that

$$\ell \leq \begin{cases} 2740 \text{ if } 612 \leq k \leq 1102 \\ 8870 \text{ if } 1103 \leq k \leq 1352. \end{cases}$$

We calculate the exact values of ℓ' for each $\ell \geq 31$ and check that inequality (6.10) is not satisfied. This completes the proof of the lemma. □

Finally we deal with the remaining case $30 \leq k \leq 7173$, $\ell = 3, 5, 7$. We show

Lemma 6.4.5 *Suppose (5.2) holds with* $\ell \in \{3, 5, 7\}$ *and* $30 \leq k \leq 7173$. *Then*

$$D_1 > \beta_0 \theta k^{\ell - \ell' - \ell'/\ell}$$

where ℓ' is

$$\ell' = \begin{cases} 2 \text{ if } \ell = 3 \\ 3 \text{ if } \ell = 5 \\ 4 \text{ if } \ell = 7. \end{cases}$$

Proof Suppose

$$D_1 \leq \beta_0 \theta k^{\ell - \ell' - \ell'/\ell}.$$

Then Lemmas 2.2.1 and 2.4.4 are valid. For $30 \leq k \leq 1352$ we use (6.11) to contradict the inequality Lemma 2.2.1(i). When $1353 \leq k \leq 7173$ we take $m_1 = 5, m_2 = 11$. We find that $F_0(k, 5, 11) \geq 259$. Using this we see that the inequality Lemma 2.2.1(ii) is not valid. This proves the assertion of the lemma. □

6.5 Proofs of Theorems 6.1.1, 6.1.2 and 6.1.3

Theorem 6.1.3 is immediate from Lemma 6.4.1. Theorem 6.1.2 is a direct consequence of combining Lemmas 6.4.2–6.4.5 together with Table 6.3. Now we go to

Proof of Theorem 6.1.1. Let $D_1 = 1$. By Theorem 6.1.2, it is easy to see that we need to consider only

$$k \leq \begin{cases} 364 \text{ if } \ell = 3 \;\&\; 3 \mid d \\ 13 \text{ if } \ell = 3 \;\&\; 3 \nmid d \\ 4 \text{ if } \ell = 5 \;\&\; 5 \mid d \\ 4, 5, 6, 7 \text{ if } \ell = 5 \;\&\; 5 \nmid d \\ 4, 5 \text{ if } \ell = 7 \;\&\; 7 \mid d. \end{cases}$$

Let us look at $\ell = 3 \;\&\; 3 \mid d$ and $k \leq 364$. Since (6.3) with $D_1 = 1$ is satisfied for $k \geq 4$, all a_i's are distinct. Further they are cube free and $3 \nmid a_i$ since $3 \mid d$. Suppose $g(k, 1)$ and $g(k, 2)$ denote the number of a_i's composed of 2 and 2, 5 respectively. Then we see that

$$g(k, 1) \leq 3;\; g(k, 2) \leq 9 \tag{6.12}$$

and

$$g(k, 1) \geq f(k, 2) \geq f_0(k, 2);\; g(k, 2) \geq f(k, 3) \geq f_0(k, 3).$$

We check that

$$f_0(k, 2) \geq 4 \text{ for } 4 \leq k \leq 40;\; f_0(k, 3) \geq 10 \text{ for } 41 \leq k \leq 102$$

which contradicts (6.12). So we may assume that $103 \leq k \leq 364$. Since we know that $3|d$, we use α_4', α_5' and α_7' with $\delta = (0.5)$ given in Exercise 6 (1) to get $\alpha_7' > 1$.

This implies that $D_1 > 1$. Some of the cases occur since we presented a uniform result in Theorem 6.1.2. By computing a lower bound for $n + (k-1)d$ for each case, we can exclude

$$\ell = 3, 7 \le k \le 13, 3 \nmid d;\ \ell = 5, 5 \nmid d, k = 7;\ \ell = 7, 7|d, k = 5.$$

For instance, for $k = 7, \ell = 3, 3 \nmid d$ we get $\alpha_3 \ge 0.5543$ and we check that $D_1 > 1$. Thus we are left with

$$k \le \begin{cases} 4, 5, 6 \text{ if } \ell = 3 \ \&\ 3 \nmid d \\ 4 \text{ if } \ell = 5 \ \&\ 5 \mid d \\ 4, 5, 6, \text{ if } \ell = 5 \ \&\ 5 \nmid d \\ 4 \text{ if } \ell = 7 \ \&\ 7 \mid d. \end{cases}$$

Let us look at $k = 5, 6$. Then $\ell = 3, 3 \nmid d;\ \ell = 5, 5 \nmid d$. So $\theta = 1$. Suppose a prime ≥ 11 divides the product Δ. Then $n + (k-1)d \ge 11^{\ell}$ so that

$$(0.7)\frac{\theta}{k}(n + (k-1)d)^{(\ell - \ell')/\ell} \ge (0.7)\frac{11}{k} > 1.$$

Thus $D_1 > 1$. So we may assume that $P(\Delta) = 7$. Then $P(A_i) \le 5$ and 7 divides only one term and so all X_i's are equal to 1 except one of them. By Erdős argument, we get

$$n(n+d) \le 120 \text{ if } k = 6;\ n \le 24 \text{ if } k = 5.$$

For all possible values of n and $d > 1$ satisfying $n(n+d) \le 120$ we check that $P(\Delta) \ne 7$. Thus $k \ne 6$. The reader might have noticed that as soon as we are able to bound n and d, it is a routine check. Thus if $n + d \le 24$ when $k = 5$ we would have been through. But this may not be true. So we can only say $n \le 24$. We may assume that $5 \nmid n$ since otherwise we would get $n + d \le 24$. Further since the length of the product is 5, we see that in fact

$$n \in \{1, 2, 3, 4, 6, 12\}.$$

(Can you justify this?) There are three terms which are composed of only 2 and 3. It is enough to consider the following possibilities.

$$P(n(n+d)(n+2d)) \le 3;\ P(n(n+2d)(n+3d)) \le 3;\ P(n(n+3d)(n+4d)) \le 3.$$

Let us consider

$$P(n(n+d)(n+2d) \le 3.$$

Then we get

$$(n, n+d, n+2d) \in \{(1, 2^{\alpha_1}, 3^{\beta_2}), (2, 3^{\beta_1}, 2^{\alpha_2})\}.$$

The reader can check easily that there is no other possibility. Thus

$$2^{\alpha_1+1} - 3^{\beta_2} = 1;\ 3^{\beta_1} - 2^{\alpha_2-1} = 1.$$

These are particular cases of the well-known *Catalan* equation. The first equation has no solution. The only possible solution of the second equation is

$$\beta_1 = 2, \alpha_2 = 4$$

which gives

$$(n, n+d, n+2d) = (2, 9, 16)$$

so that

$$(n, n+d, n+2d, n+3d, n+4d) = (2, 9, 16, 23, 30).$$

and we see that $P(\Delta) \neq 7$. The other two possibilities can be excluded similarly. Thus $k \neq 5$.

Finally we are left with $k = 4$. First we see that $D_1 > 1$ except when

$$(i)\ \ell = 3, 3 \nmid d;\ \ell = 5, 5 \nmid d;\ \ell = 7, 7 | d \text{ with } P(\Delta) = 5$$

and

$$(ii)\ \ell = 5, 5 | d \text{ with } 7 \leq P(\Delta) \leq 23.$$

In the case (i) we use Erdős argument to get a term

$$n + id \leq 6.$$

If $i \neq 0$, we check that $P(\Delta) \neq 5$. Suppose $i = 0$, i.e., $n \leq 6$. In fact

$$n \in \{1, 2, 3, 6\}.$$

Also we have the following possibilities.

$$(P(n(n+d)(n+2d)) \leq 3;\ P(n(n+2d)(n+3d)) \leq 3.$$

We argue as earlier to rule out these possibilities. Now we take the latter case (ii). (Since $k = 4$, the fact that $5|d$ has no impact in Lemma 6.3.1 unlike the case $3|d$ which we came across earlier). We may assume that each X_i is divisible by at most one prime from $\{7, 11, 13, 17, 19, 23\}$. Otherwise $D_1 > 1$. By removing the A_i's in which 2 and 3 appear to the maximum, we get an A_i and A_j such that

$$A_j A_j \leq 6.$$

Thus

$$(A_i, A_j) \in \{(1, 1), (1, 2), (1, 3), (1, 4), (1, 6), (2, 2), (2, 3)\};$$

$$(i - j)d = A_i X_i^5 - A_j X_j^5$$

and hence the right-hand side is divisible by 5 since $5|d$. If X_i or X_j equals 1 then $n \leq 6$ and this can be excluded as earlier. So we have

$$X_i, X_j \in \{7, 11, 13, 17, 19, 23\}.$$

Let Z be the set of tuples $(A, X; B, Y)$ with (A, B) taking the possible values of A_i, A_j and X, Y taking the possible values of (X_i, X_j) such that $AX^5 - BY^5 > 0$ and divisible by 5. We know that $i - j = 1, 2, 3$. Suppose $i - j = 1$, then $(i, j) \in \{(1, 0), (2, 1), (3, 2)\}$. Consider $(i, j) = (1, 0)$. Then

$$d = A_1 X_1^5 - A_0 X_0^5;\ A_2 X_2^5 = 2A_1 X_1^5 - A_0 X_0^5;\ A_3 X_3^5 = 3A_1 X_1^5 - 2A_0 X_0^5;$$

$$P(A_2 X_2^5) \leq 23,\ P(A_3 X_3^5) \leq 23.$$

Let us describe the procedure. Take any $(A, X; B, Y)$ from the set Z. In the present case, $(A, X; B, Y) = (A_1, X_1; A_0, X_0)$. We check first that

$$P(2AX^5 - BY^5) \leq 23;\ P(3AX^5 - 2BY^5) \leq 23.$$

If so, we check further if every prime >4 occurs to 5th power in $2AX^5 - BY^5$ and/or in $3AX^5 - 2BY^5$. We carry out this procedure for every tuple in Z. We found no tuple satisfying the required conditions. So we conclude that $(i, j) \neq (1, 0)$. In the case $(i, j) = (2, 1)$, we follow the procedure for $P(2BY^5 - AX^5)$ and $P(2AX^5 - BY^5)$ and if $(i, j) = (3, 2)$ then we check with $P(2BY^5 - AX^5)$ and $P(3BY^5 - 2AX^5)$ for every $(A, X; B, Y) \in Z$.

Next, let $i - j = 2$. Then $(i, j) \in \{(2, 0), (3, 1)\}$. Let Z_1 be the subset of Z of those tuples $(A, X; B, Y)$ with $10|(AX^5 - BY^5)$. Here we follow the procedure for $P(AX^5 + BY^5)$, $P(3AX^5 - BY^5)$, $P(3BY^5 - AX^5)$ with $(A, X; B, Y) \in Z_1$.

Finally, let $i - j = 3$. Then $(i, j) = (3, 0)$. Let Z_2 be the subset of Z of those tuples $(A, X; B, Y)$ with $15|(AX^5 - BY^5)$. We follow the procedure for $P(AX^5 + 2BY^5)$, $P(2AX^5 + BY^5)$ with $(A, X; B, Y) \in Z_2$.

By this verification, all possibilities are excluded. This shows that $k \neq 4$. □

6.6 Comments

Theorem 6.1.2 is better than the result in [3, Theorem 3]. This is mainly due to more precise computations and the improvement in the graph theoretic argument in Lemmas 6.4.3 and 6.4.4. There will be further improvement if we know that d is divisible by small primes like 2, 3 and 5. This was exploited in [4]. In fact the first improvement in the graph theoretic argument also occurred in the same paper. For instance, it is shown that if d is divisible by 2 or 3, then (5.2) with $k \geq 4$ and $\ell \geq 5$ implies that

$$D_1 > (0.7)\theta k^{\ell-3}. \tag{6.13}$$

The results of this chapter give a lower bound for d. In [5], Saradha and Shorey showed that in fact (5.2) does not hold for large values of d which is much larger for small values of k and ℓ. For instance, it was shown that for $\ell \geq 11, d > 10^{15}$ whereas from Table 6.1, we could get only $d > 0.93\theta k^{25/11}$. Of course, as k and ℓ become larger, Theorem 6.1.2 becomes stronger. The reader is encouraged to look into their paper.

Exercise 6

(1) Suppose there is a prime $p < k$ which divides d. Then show that α_4, α_5 and α_7 in Lemma 6.3.1 can be replaced by α_4', α_5' and α_7' where

$$\alpha_4' = \alpha_4 + \frac{1}{k} + \frac{\log p}{(p-1)(\ell-1)\log k},$$

$$\alpha_5' = \alpha_5 + \frac{1}{k} + \frac{\log p}{(p-1)(\ell \log k + \log \alpha_4')}$$

and

$$\alpha_7' = \left(\alpha_5' - \delta + \frac{\pi(k)}{k}\right)(\delta \log(\delta k))^{\ell},$$

respectively.
(Remember, this is used in the proof of Theorem 6.1.1 when $\ell = 3, 3|d$.)

(2) Write computer algorithms and verify all the calculations mentioned in the chapter.

(3) Follow the methods of Chap. 6 and show (6.13).

References

1. T.N. Shorey, Some exponential Diophantine equations, in *New Advances in Transcendence Theory* ed. by A. Baker (Cambridge University Press, 1988), pp. 352–365. [**6**]
2. T.N. Shorey, R. Tijdeman, Perfect powers in products of terms in an arithmetical progression. Compositio Math. **75**, 307–344 (1990). [**4, 5, 6**]
3. N. Saradha, T.N. Shorey, Almost perfect powers in arithmetic progression. Acta Arith. **138**, 363–388 (2001). [**6, 7**]
4. N. Saradha, R. Tijdeman, Arithmetic progressions with common difference divisible by small primes. Acta Arith. **131**, 267–279 (2008). [**2, 6**]
5. N. Saradha and T. N. Shorey, *Contributions towards a conjecture of Erdős on perfect powers in arithmetic progression,* Compositio Mathematicae **141** (2005), 541-560. [**6**]

Chapter 7
Generalized Fermat Equations and the Conjecture of Erdős

$\frac{4!}{3} = 2^3; 9262^3 + 15312283^2 = 113^7$

In 1637, the Frenchman Pierre de Fermat stated that the equation

$$x^\ell + y^\ell = z^\ell$$

does not have any solution in non-zero integers for any integer $\ell > 2$. When $\ell = 2$, we get Pythagorean equation and it is well known that there are infinitely many integral solutions. The above equation is a particular case of the *Generalized Fermat Equation* (GFE)

$$Ax^\ell + By^\ell = Cz^\ell, A, B, C \in \mathbb{Z}. \tag{7.1}$$

Fortunately, Fermat never wrote down the proof and many beautiful branches of number theory have been discovered in an attempt to find a proof of it. In the nineteenth century, Kummer developed the idea of algebraic number fields but, not successful in solving Fermat conjecture. In 1966, Siegel [1] showed a fundamental result that

the number of integral points on an irreducible algebraic curve of genus > 0 *is finite.*

After this, the breakthrough ideas came in 1980s. In 1983, Faltings [2] proved an extraordinary theorem on Mordell's conjecture, by showing that

the number of rational points on an irreducible algebraic curve of genus > 1 *is finite.*

As a consequence, one gets

Let $\ell \geq 4$. *Then there are only finitely many coprime integers satisfying* (7.1).

These results are *ineffective.* After Faltings' result, many spectacular advances were made by Ribet and many other mathematicians, culminating in the works of Wiles

S. Natarajan, *Perfect Powers—An Ode to Erdős*, Infosys Science Foundation Series,
https://doi.org/10.1007/978-981-96-2599-4_7

in 1994 when he solved Fermat's last theorem. The reader may be wondering why suddenly we are shifting to GFE. Let us look at Eq. (5.2) when $k = 3$. We have not said anything about this case in Chap. 6. Consider

$$n(n+d)(n+2d) = by^{\ell},\ P(b) \le 2,\ \ell \ge 3,\ \gcd(n,d) = 1.$$

We write

$$n = a_0 x_0^{\ell},\ n+d = a_1 x_1^{\ell},\ n+2d = a_2 x_2^{\ell},\ P(a_i) \le 2.$$

By gcd conditions, it is easy to see that

$$(a_0, a_1, a_2) \in \{(1,1,1), (1, 2^{\alpha_1}, 1), (2, 1, 2^{\alpha_2}), (2^{\alpha_0}, 1, 2)\}.$$

Using the identity

$$n + (n+2d) = 2(n+d)$$

and ℓ odd we get an equation of the form $x^{\ell} + y^{\ell} = 2^{\alpha} z^{\ell}$ which has no non-trivial solution by Theorem 1.6.15. Trivial solutions do not give rise to any solution to our equation. Thus we conclude that

Equation (5.2) *with* $k = 3$ *and* $P(b) \le 2$ *has no solution.*

This was first observed by Győry [3]. It indicates that if we could form *ternary* equations as in (7.1), it may be possible to use results from Fermat type equations. Forming ternary equations is not difficult. If $0 \le p < q < r < k$, then we have the identity

$$(r-q)(n+pd) + (q-p)(n+rd) = (r-p)(n+qd)$$

which gives

$$(r-q)a_p x_p^{\ell} + (q-p)a_r x_r^{\ell} = (r-p)a_q x_q^{\ell}.$$

We shall call this identity as $[p, q, r]$. This is a ternary equation of *signature* (ℓ, ℓ, ℓ). The problem is to get control of the coefficients. Since the coefficients depend on k it may be possible to bound the coefficients if k is small. Then perhaps, we can treat such equations with methods used in treating some cases of (7.1). Here is where the theory of modular forms and elliptic curves enter. This started in a simple way in the papers of Sander [4], Saradha and Shorey [5, Lemma 13]. Bennett and Skinner [6] and later Bennett et al. [7], Győry et al. [8] and Győry et al. [9] have made important contributions in this direction. This helped in solving (5.2) for small values of k with some restriction on $P(b)$. It is now known (see [9, Theorem 1.1] that

Equation (5.2) *with* $3 \le k < 35$ *and* $b = 1$ *(i.e., the case of perfect powers) has no solution.*

Thus we see a blend of modern and classical methods in dealing with the problem of Erdős on perfect powers in products. In [5], it was shown that

$8(=2^3)$ *is the only perfect power that can be expressed as a product of* $k \geq 3$ *consecutive positive integers with one term omitted.*

This together with

$$\frac{6!}{5} = 12^2, \frac{10!}{7} = 720^2$$

(see Sect. 4.5) gives a complete answer to a conjecture of Erdős and Selfridge. Let $\Delta' = \Delta'(n, k, 1)$ denote the product $\Delta = \Delta(n, k, 1)$ with one term omitted. To give the reader an appetizer of the ideas coming from Generalized Fermat Equations and how they merge with Erdős classical ideas, we have chosen the product Δ' with $k = 6$ (Theorem 7.1.1). As a main course, we have taken a more involved product $\Delta(n, d, k)$ with $k \leq 8$ (Theorem 7.1.2).

7.1 Statement of the Results

The following theorem is a particular case of ([5, Theorem 1, p. 386]).

Theorem 7.1.1 *The equation*

$$\Delta' = \Delta'(n, 6, 1) = y^\ell, \ell \geq 3, \ell \ prime$$

has no solution for $n > 0$.

The next result is taken from [7, Theorem 1.2].

Theorem 7.1.2 *Let* $d > 1$. *The equation*

$$\Delta(n, k, d) = by^\ell, 3 \leq k \leq 8, P(b) \leq 2, \ell \geq 7, \ell \ prime$$

has no solution for $n > 0$.

7.2 A Result on GFE and the Proof of Theorem 7.1.1

We refer to Sect. 1.5 for all the terminologies. Since $k \leq 8$ in both the theorems, a_i's are composed of only the primes 2 or 3 or 5 or 7. So we shall use the following notation

$$a_i = 2^{\alpha_i} 3^{\beta_i} 5^{\gamma_i} 7^{\delta_i}, 0 \leq i < k$$

with $\alpha_i, \beta_i, \gamma_i, \delta_i$ integers ≥ 0. We put

$$\Delta_a = \prod_{i=0}^{k-1} a_i.$$

Note that by the given equations in the theorems,

$$\sum \alpha_i \equiv \nu_2(b) \pmod{\ell}; \sum \beta_i \equiv \sum \gamma_i \equiv \sum \delta_i \equiv 0 \pmod{\ell}.$$

We use gcd conditions between the terms in the product to position these small primes and determine their exponents.

Proof of Theorem 7.1.1. We have the equation

$$\Delta'(n, 6, 1) = y^\ell, \ell \geq 3.$$

Since we know already that product of five consecutive terms is never a cube or higher power, we may assume that the omitted term is say, $n + i_0$ with $i_0 \in \{1, 2, 3, 4\}$. By mirror image or symmetry, it is enough to consider $i_0 \in \{3, 4\}$ which give the following two equations.

$$(i)\ n(n+1)(n+2)(n+4)(n+5) = y^\ell$$

and

$$(ii)\ n(n+1)(n+2)(n+3)(n+5) = y^\ell.$$

Suppose $P(\Delta'(n, 6, 1)) \leq 5$. Then by Erdős argument, there exist at least two terms $n + i, n + j$ such that

$$n(n+1) \leq (n+i)(n+j) \leq 5!$$

giving $n \leq 10$. One can check easily that for $1 \leq n \leq 10$, $P(\Delta'(n, 6, 1)) > k(= 6)$ except when

$$i_0 = 3 : n = 1, 4\ ;\ i_0 = 4 : n = 1, 3.$$

In these cases we see that Δ' is not a perfect power. So we may assume that

$$P(\Delta'(n, 6, 1)) \geq 7.$$

Thus Lemma 2.4.3 holds (Check if omitting a term makes any difference). In particular all a_i's are distinct. So $5 \geq f(k, 2) \geq 3$. Further by Lemma 2.2.1

$$\binom{f(k, 2) + \ell' - 1}{\ell'} \leq \ell^2.$$

With $\ell' = \ell - 1$ this gives

$$\binom{f(k, 2) + \ell - 2}{\ell - 1} \leq \ell^2.$$

This is not satisfied if $f(k, 2) = 4, 5$. So we get $f(k, 2) = 3$. This means the prime 5 divides exactly two a_i's which are a_0 and a_5.

Let us consider equation (i).

Case 1. *Suppose some* a_i *equals 1.* Let $a_{i_0} = 1$. Then

$$\prod_{i \neq i_0} a_i = a_{i_0}^{\ell-1} t^{\ell}$$

where t is an integer. This is not possible by Lemma 2.4.3 with $\ell' = \ell - 1$.

Case 2. *Suppose no* a_i *equals 1.* Then a simple analysis (try!) shows that $6|(n+4)$ and so

$$a_0 = 2^{\alpha_0} 5^{\gamma_0};\ a_1 = 3^{\beta_1};\ a_2 = 2^{\alpha_2};$$

$$a_4 = 2^{\alpha_4} 3^{\beta_4};\ a_5 = 5^{\gamma_5}$$

with

$$\alpha_0 + \alpha_2 + \alpha_4 \equiv 0 \pmod{\ell}, \alpha_2 > 0;\ \beta_1 + \beta_4 \equiv 0 \pmod{\ell}, \beta_1 > 0;$$

$$\gamma_0 + \gamma_5 \equiv 0 \pmod{\ell}, \gamma_5 > 0.$$

Since a_i's are ℓth power free and $b = 1$ we get that

$$(\alpha_0, \alpha_2, \alpha_4) \in \{(\ell - 3, 1, 2), (1, \ell - 2, 1), (2, 1, \ell - 3)\};$$

$$(\beta_1, \beta_4) \in \{(1, \ell - 1), (\ell - 1, 1)\}; (\gamma_0, \gamma_5) \in \{(1, \ell - 1), (\ell - 1, 1)\}.$$

Let $(\alpha_0, \alpha_2, \alpha_4)$ equal either $(\ell - 3, 1, 2)$ or $(2, 1, \ell - 3)\}$. Then we find that either

$$a_1 a_4 = a_2^2 3^{t\ell} \text{ or } a_0 a_5 = a_2^2 5^{t\ell} \text{ for some integer } t > 0.$$

This contradicts Lemma 2.4.3 with $\ell' = 2$. Next let $(\alpha_0, \alpha_2, \alpha_4) = (1, \ell - 2, 1)$. Form identity $[p, q, r] = [1, 2, 4]$ to get

$$2 \times 3^{\beta_1} x_1^{\ell} + 2^{\alpha_4} 3^{\beta_4} x_4^{\ell} = 3 \times 2^{\alpha_2} x_2^{\ell}$$

which reduces to

$$3^{\beta_1 - 1} x_1^{\ell} + 3^{\beta_4 - 1} x_4^{\ell} = 2^{\ell - 3} x_2^{\ell}.$$

By Lemma 1.6.14 there is no non-zero solution to this equation whenever $\ell \geq 7$. Thus we need to consider $\ell = 3, 5$. Let $\ell = 5$. Then $\alpha_2 = 3$ and

$$a_0 a_1 a_4 a_5 = 2^{\alpha_0 + \alpha_4} 3^5 5^5 = 2^2 3^5 5^5$$

and

$$a_2^4 = 2^{12} = 2^2 \times 2^{10}.$$

Thus

$$a_0 a_1 a_4 a_5 = a_2^4 \left(\frac{15}{4}\right)^5$$

which contradicts Lemma 2.4.3 with $\ell' = \ell - 1 = 4$. Thus $\ell \neq 5$. Let $\ell = 3$. Then $(\alpha_0, \alpha_2, \alpha_4) = (1, 1, 1)$ and

$$a_0 a_5 = a_1 a_4 \left(\frac{5}{3}\right)^3$$

which contradicts Lemma 2.4.3 with $\ell' = 2$. Thus equation (i) does not hold.

Consider equation (ii)

$$n(n+1)(n+2)(n+3)(n+5) = y^\ell.$$

As earlier, we need to consider the case when no a_i equals 1. Then we see that

$$a_0 = 5^{\gamma_0}, a_1 = 2^{\alpha_1}, a_2 = 3^{\beta_2}, a_3 = 2^{\alpha_3}, a_5 = 2^{\alpha_5} 3^{\beta_5} 5^{\gamma_5}$$

with $\alpha_1, \alpha_3, \beta_2, \gamma_0$ positive integers and

$$(\alpha_1, \alpha_3, \alpha_5) \in \{(\ell - 3, 1, 2), (1, \ell - 2, 1), (2, 1, \ell - 3)\}$$

$$(\beta_2, \beta_5) \in \{(1, \ell - 1), (\ell - 1, 1)\}; (\gamma_0, \gamma_5) \in \{(1, \ell - 1), (\ell - 1, 1)\}.$$

Suppose $(\alpha_1, \alpha_3, \alpha_5) = (2, 1, \ell - 3)$. Then $|x_3^\ell - 2x_1^\ell| = 1$ which by Lemma 1.2.6 gives $x_1 = x_3 = 1$. This gives $n + 1 = 4$, i.e., $n = 3 \neq 5^{\gamma_0}$. Now we assume that $(\alpha_1, \alpha_3, \alpha_5) = (\ell - 3, 1, 2)$ or $(1, \ell - 2, 1)$. Then consider the identity $[p, q, r] = [1, 2, 3]$ which is

$$a_1 x_1^\ell + a_3 x_3^\ell = 2 a_2 x_2^\ell.$$

This reduces to

$$(a)\ 2^{\ell-4} x_1^\ell + x_3^\ell = 3^{\beta_2} x_2^\ell$$

or

$$(b)\ x_1^\ell + 2^{\ell-3} x_3^\ell = 3^{\beta_2} x_2^\ell.$$

By Lemma 1.6.14 there is no non-zero solution to this equation whenever $\ell \geq 11$ if (a) holds and for $\ell \geq 7$ if (b) holds. So we may assume that

$$\ell \leq 7 \text{ if } (\alpha_1, \alpha_3, \alpha_5) = (\ell - 3, 1, 2)$$

and

$$\ell \le 5 \text{ if } (\alpha_1, \alpha_3, \alpha_5) = (1, \ell - 2, 1).$$

Let $(\alpha_1, \alpha_3, \alpha_5) = (\ell - 3, 1, 2)$ and $\ell \le 7$. Since no a_i equals 1, we get $\ell \ne 3$. Further

$$a_0 a_1 a_2 a_5 = a_3^4 (15)^5 \text{ if } \ell = 5$$

and

$$a_0 a_2 a_5 = a_1^2 a_3 \left(\frac{15}{2}\right)^{\ell} \text{ if } \ell = 7$$

a contradiction to Lemma 2.4.3 with $\ell' = 4$ and 3, respectively. Let $(\alpha_1, \alpha_3, \alpha_5) = (1, \ell - 2, 1)$, $\ell \le 5$. Since a_i's are not equal, we get $\ell \ne 3$. When $\ell = 5$,

$$a_0 a_2 a_3 a_5 = a_1^4 (15)^{\ell},$$

a contradiction to Lemma 2.4.3 with $\ell' = 4$. □

Note: The results on GFE are used to bound ℓ. To exclude small values of ℓ, combinatorial ideas of Erdős come handy.

7.3 Proof of Theorem 7.1.2

Here we do not have the facility of using Lemma 2.4.3 since we are not having any restriction on d. We begin with a simple congruence criterion.

Lemma 7.3.1 *Let*

$$ax^7 + by^7 = cz^7$$

with $7^2 | ax^7$ *and* $7 \nmid bc(yz)^7$. *Then*

$$bc^{-1} \equiv \pm 1, \pm 18, \pm 19 \pmod{7^2}.$$

Proof By the given hypothesis,

$$by^7 \equiv cz^7 \pmod{7^2}$$

so that, upon taking sixth powers, it follows that

$$b^6 \equiv c^6 \pmod{7^2}.$$

(Why?) Thus

$$(bc^{-1})^6 \equiv 1 \pmod{7^2}$$

implying the assertion of the lemma. □

Besides the identity $[p, q, r]$, on many occasions we use the following identity. If p, q, r, s are integers with

$$0 \leq p < q \leq r < s \leq k-1 \text{ and } p+s = q+r$$

then we have the identity

$$(n+qd)(n+rd) - (n+pd)(n+sd) = (qr - ps)d^2 \neq 0.$$

We shall call this identity as $[p, q, r, s]$. This leads to a ternary equation of the form

$$Ax^\ell + By^\ell = Cz^2 \text{ with } P(AB) < k.$$

This is a ternary equation of *signature* $(\ell, \ell, 2)$. In the next lemma, we shall give results on some such ternary equations which we require for the proof of Theorem 7.1.2. The reader may already have an idea about the proofs of such results from Sect. 1.6. On a more complicated level, the following lemma was proved in [6] and [7, Proposition 3.1].

Lemma 7.3.2 *Let $\ell \geq 7$ be prime, α and β be non-negative integers, and let A and B be coprime non-zero integers with A odd. Then the following equations have no solutions in non-zero coprime integers (x, y, z) with $xy \neq \pm 1$.*

(i) $x^\ell + 2^\alpha y^\ell = 3^\beta z^2, \alpha \neq 1,$
(ii) $x^\ell + y^\ell = Dz^2, D \in \{2, 6\},$
(iii) $x^\ell + 2^\alpha y^\ell = z^2 \text{ with } p|xy \text{ for } p \in \{3, 5, 7\},$
(iv) $Ax^\ell + By^\ell = z^2, P(AB) \leq 3 \text{ with } p|xy \text{ for } p \in \{5, 7\},$
(v) $x^\ell + 2^\alpha y^\ell = 3z^2 \text{ with } p|xy \text{ for } p \in \{5, 7\},$
(vi) $x^\ell + 5^\alpha y^\ell = 2z^2, \ell \geq 11, \alpha > 0,$
(vii) $Ax^\ell + By^\ell = z^2, P(AB) \leq 5, 7|xy, \ell \geq 11.$

Proof We refer to Sect. 1.6 for all the terminologies and results on Modular forms. We will be giving only the main argument involved. For complete details we refer to [6, 7]. Let us assume that given an equation from *(i)*–*(vii)*, there exists a non-trivial solution (x, y, z). Let us write every equation in the form

$$a + b = c$$

where $a = Ax^\ell, b = By^\ell$ and $c = Cz^\ell$. Note that exactly one of a, b, c is even. Assume without loss of generality that a and c are odd and b is even. Further assume that

$$a \equiv -1 \pmod 4.$$

Associate the Frey curve $E/\mathbb{Q}$

$$Y^2 = X(X - a)(X + b).$$

Let N_E and Δ_E be the conductor and minimal discriminant of E. Suppose E has no ℓ-isogeny. Then there is a newform f of weight 2 and trivial Nebentypus character with N_ℓ as its level and such that

$$E \sim_\ell f.$$

Let the newform f have Fourier coefficients a_n in a Number field K_f. As an example, let us consider the equation

$$x^\ell + 2y^\ell = z^2 \text{ with } p|xy \text{ for } p \in \{3, 5, 7\}.$$

This is given in equation (*iii*) with $\alpha = 1$. Let us assume that there exists a non-trivial solution (x, y, z) with $\gcd(x, y, z) = 1$. The level of the associated newform is $N = N_\ell = 128$. When $p|xy$ and if p is coprime to ℓN, then

$$\text{Norm}_{K_f/\mathbb{Q}}(a_p \pm (p + 1)) \equiv 0 \pmod \ell \tag{7.2}$$

with $|a_p| \leq 2\sqrt{p}$. Then one can get the value of a_p for each p in the newform at the level N_ℓ from William Stein's 'Modular Forms Database'. Let us take $p = 3, 5$ and 7. Then we get

$$a_3 \in \{0, \pm 2\}, a_5 \in \{0, \pm 2, \pm 4\} \text{ and } a_7 \in \{0, \pm 2, \pm 4\}.$$

This contradicts (7.2) since $\ell \geq 7$. Hence (*ii*) with $\alpha = 1$ is not possible. There are occasions when $K_f \neq \mathbb{Q}$. For instance, let us consider equation (*vii*). From Table 7.1, we see that $N_\ell = 2^7 5$ when $P(AB) = 10$. In this case, $K_f = \mathbb{Q}$ or $\mathbb{Q}(\sqrt{5})$. When $K_f = \mathbb{Q}$, we have $(p, a_p) = (7, 0), (7, \pm 2)$. Then

$$\text{Norm}(a_p \pm (p + 1)) = 64, 100, 36 \equiv 0 \pmod \ell$$

which are not possible since $\ell \geq 11$. Let now $K_f = \mathbb{Q}(\sqrt{5})$. Suppose $\ell \neq 11, 19$. Then take $(p, a_p) = (7, \pm(1 + \sqrt{5}))$ so that

$$\text{Norm}(\pm(7 + 1) \pm (1 + \sqrt{5})) = 76, 44.$$

Table 7.1 Data for equations (*i*)–(*vii*) in Lemma 7.3.2

Equations no.	Δ'_E	N'_E	N_ℓ	K_f	(p, a_p)
(i)	$2^{2\alpha+\delta}3^{3\beta}$,	–	–	–	–
$\alpha = 0$	$\delta \in \{6, -12\}$	$2^\gamma 3^{2\beta}$	$2^\gamma 3^{2\beta}$	–	–
$\alpha \geq 2$	–	$2^{\gamma+1}3^{2\beta}$	$2^{\gamma+1}3^{2\beta}$	–	–
		$-1 \leq \gamma \leq 5$	$0 \leq \gamma \leq 5$	–	–
		$\gamma \neq 3$	$\gamma \neq 4$		
(ii)					
$D = 2$	2^9	2^8	2^8	–	–
$D = 6$	$2^9 3^3$	$2^8 3^2$	$2^8 3^2$	–	–
(iii)					
$\alpha = 0$	2^6	2^5	2^5	$\mathbb{Q}$	$(5, -2)$
$\alpha = 1$	2^8	2^7	2^7	$\mathbb{Q}$	$(5, \pm 2)$
$\alpha = 2$	2^4	$2^2, 2^3$	$2^2, 2^3$	No newform	–
$\alpha \in \{3, 4, 5\}$	$2^{2\alpha}$	$2^3, 2^5$	2^5	$\mathbb{Q}$	$(5, -2)$
			2^3	No newform	–
$\alpha \geq 6$	$2^{2\alpha-12}$	$1, 2$	$1,2$	No newform	–
(iv)	$2^\delta AB^2$	$2^\gamma \prod_{p\mid AB} p$			
	$\delta \in \{6, 0, -12\}$	$-1 \leq \gamma \leq 6$			
		$\gamma \neq 3$			
$P(AB) \leq 2$			2^γ	$\mathbb{Q}$	As in (*iii*)
$P(AB) = 3$			$2^\gamma 3$	$\mathbb{Q}$	$(5, 0)$
			$0 \leq \gamma \leq 7$		$(5, \pm 2)$
			$\gamma \neq 4$		$(5, \pm 4)$
(v)					
$\alpha = 0$	$2^6 3^3$	$2^5 3^3$	$2^5 3^3$	$\mathbb{Q}$	$(5, 0)$
$\alpha \geq 1$	$2^{\delta+2\alpha}3^3$	$2^\gamma 3^2$	$2^\gamma 3^2$	$\mathbb{Q}$	$(5, \pm 2)$
–	$\delta \in \{6, 0, -12\}$	$0 \leq \gamma \leq 7$	$0 \leq \gamma \leq 7$		$(5, \pm 4)$
		$\gamma \neq 4$	$\gamma \neq 4$		
(vi)	$2^9 5^{2\alpha}$	$2^8 5$	$2^8 5$	$\mathbb{Q}[\sqrt{2}]$	$(3, \sqrt{2})$
				$\mathbb{Q}[\sqrt{10}]$	$(3, \sqrt{10})$
				$\mathbb{Q}[\sqrt{3}]$	$(3, \pm 1 + \sqrt{3})$
(vii)	$2^\delta AB^2$	$2^\gamma \prod_{p\mid AB} p$	$2^\gamma, 2^\gamma 3$	$\mathbb{Q}$	$(7, 0)$
	$\delta \in \{6, 0, -12\}$	$-1 \leq \gamma \leq 5$	$2^\gamma 5$	$\mathbb{Q}$	$(7, \pm 2)$
		$\gamma \neq 3$	$2^\gamma 15$	$\mathbb{Q}$	$(7, \pm 4)$
			$0 \leq \gamma \leq 7$	$\mathbb{Q}[\sqrt{2}]$	$(7, 2\sqrt{2})$
			$\gamma \neq 4$		
–	–	$2^6 \prod_{p\mid AB} p$	$2^7 15$	$\mathbb{Q}[\sqrt{17}]$	$(7, \pm(1 + \sqrt{17})$
–	–	$2^6 \prod_{p\mid AB} p$	$2^7 5$	$\mathbb{Q}[\sqrt{5}]$	$(7, \pm(1 + \sqrt{5})$
					if $\ell \neq 11, 19$
				$\mathbb{Q}[\sqrt{5}]$	$(3, \pm(1 + \sqrt{5})$
					if $\ell = 11, 19$

This yields a contradiction since $\ell \neq 11, 19$. In the two cases $\ell = 11, 19$, take $(p, a_p) = (3, \pm(1+\sqrt{5}))$. Then

$$\mathrm{Norm}(\pm(3+1) \pm (1+\sqrt{5})) = 20, 4$$

which yields the necessary contradiction. In all the cases (*i*)–(*vii*), we can write

$$\Delta_E = \Delta'_E (xy^2)^\ell; \, N_E = N'_E \prod_{p|xy} p.$$

In Table 7.1 we provide the values of Δ'_E, N'_E, the Serre conductor N_ℓ, the field of coefficients K_f, the prime p and the value of a_p. The reader can derive the contradiction as in the examples given above for the values of ℓ specified in the lemma. The proof for equations (*i*) and (*ii*) is different. We refer to [6, Proposition 4.3–4.4, Sect. 5] for all the details. We provide only Δ'_E, N'_E and N_ℓ for these two equations. □

Proof of Theorem 7.1.2. For a given k, first we locate the primes dividing the a_i's. Then we form a suitable equation using the identity $[p, q, r, s]$ which reduces to one of the equations given in Lemma 7.3.2. These equations have no solutions with $xy \neq 1$. This enables us to conclude that the given equation has no solution. Suppose there are three consecutive a_i's, say a_i, a_{i+1} and a_{i+2} such that

$$P(a_i a_{i+1} a_{i+2}) \leq 2.$$

Then

$$a_i x_i^\ell + a_{i+2} x_{i+2}^\ell = 2 a_{i+1} x_{i+1}^\ell$$

reduces to an equation as in Theorem 1.6.15 which has no non-trivial solution. So we may assume always that

$$P(a_i a_{i+1} a_{i+2}) \geq 3 \text{ for any } i \text{ with } 0 \leq i \leq k-3. \tag{7.3}$$

Thus $k \neq 3$ and $P(\Delta_a) \geq 3$ for $k \geq 4$. Since $P(b) \leq 2$, if any prime p, $3 \leq p < k$ divides Δ_a, then p divides at least two a_i's. Suppose 3 divides exactly two a_i's, say a_i, a_{i+3} so that $\beta_i + \beta_{i+3} \equiv 0 \pmod{\ell}$. Suppose further that $P(a_i a_{i+1} a_{i+2} a_{i+3}) = 3$. Consider $[p, q, r, s] = [i, i+1, i+2, i+3]$. This reduces to either

$$2^{\alpha_{i+1}-1} X^\ell - 2^{\alpha_{i+3}-1} Y^\ell = d^2, 3|XY$$

or

$$2^{\alpha_{i+2}-1} X^\ell - 2^{\alpha_i - 1} Y^\ell = d^2, 3|XY$$

or

$$X^\ell - Y^\ell = 2d^2$$

which are ruled out by equations Lemma 7.3.2 (*ii*) and (*iii*). So we conclude that $k \neq 4, 5$ and if $k = 6$, then $P(\Delta_a) = 5$. In the case of $k = 7, 8$ we need to investigate also the case when 3 divides three a_i's, say a_i, a_{i+3} and a_{i+6} and $P(a_i a_{i+1} a_{i+2} a_{i+3} a_{i+4} a_{i+5}\, a_{i+6}) = 3$. Then consider $[p, q, r, s] = [i + 1, i + 2, i + 4, i + 5]$. This gives rise to an equation of the form

$$x^\ell + 2^\alpha y^\ell = 3d^2 \text{ with } \alpha \neq 1$$

if $2|a_{i+1}, a_{i+5}$ or if $2|a_{i+2}, a_{i+4}$ and $\alpha_{i+2} + \alpha_{i+4} \not\equiv 1 \pmod{\ell}$. This has no solution by Lemma 7.3.2 (*i*). If $\alpha_{i+2} + \alpha_{i+4} \equiv 1 \pmod{\ell}$, then either

$$\nu_2(n + (i + 2)d) \equiv 0 \pmod{\ell} \text{ and } \nu_2(n + (i + 4)d) = \alpha_{i+4} = 1;$$

or

$$\nu_2(n + (i + 2)d) = \alpha_{i+2} = 1 \text{ and } \nu_2(n + (i + 4)d) \equiv 0 \pmod{\ell}.$$

Then we take the equation $[i + 2, i + 3, i + 4]$. This gives rise to

$$x^\ell + 2^{\ell-1} y^\ell = 3^\beta z^\ell.$$

This equation has no solution by Theorem 1.6.14 since $\ell \geq 7$. Thus we conclude that if $k = 7$ and 8, then $P(\Delta_a) \geq 5$.

Suppose $P(\Delta_a) = 5$. Since $k \leq 8$, prime 5 divides exactly two a_i's, say

$$5|a_i a_{i+5}. \tag{7.4}$$

Then $P(a_{i+1} a_{i+2} a_{i+3} a_{i+4}) = 3$. Now

$$(i)\ 3\big|a_i, a_{i+3} \text{ or } (ii)\ 3|a_{i+2} a_{i+5} \text{ or } (iii)\ 3|a_{i+1} a_{i+4}.$$

Consider the cases (*i*) and (*ii*). Take $[p, q, r, s] = [i, i + 2, i + 3, i + 5]$. We get

$$a_{i+2} a_{i+3} (x_{i+2} x_{i+3})^\ell - a_i a_{i+5} (x_i x_{i+5})^\ell = 6d^2.$$

If $2|a_i a_{i+2}$ or $2|a_{i+3} a_{i+5}$, the above equation reduces to

$$Ax^\ell + By^\ell = d^2,\ P(AB) \leq 3,\ 5|xy$$

which has no solution by Lemma 7.3.2 (*iv*). So let us assume that $2 \nmid a_i a_{i+2} a_{i+3} a_{i+5}$. Then consider $[p, q, r, s] = [i, i + 1, i + 4, i + 5]$. We get

$$a_{i+1} a_{i+4} (x_{i+1} x_{i+4})^\ell - a_i a_{i+5} (x_i x_{i+5})^\ell = 4d^2 = (2d)^2.$$

This is of the form

$$x^{\ell}+By^{\ell}=(2d)^2,\ P(B)=3,5|xy$$

which has no solution by Lemma 7.3.2 (*iv*). Next consider case (*iii*). Take

$$[p,q,r,s]=[i,i+2,i+3,i+5] \text{ if } 2\mid a_ia_{i+1}a_{i+4}a_{i+5}$$

and

$$[p,q,r,s]=[i,i+1,i+4,i+5] \text{ if } 2\nmid a_ia_{i+1}a_{i+4}a_{i+5}.$$

In the first possibility, we get an equation of the form

$$x^{\ell}+2^{\alpha}y^{\ell}=3d^2,\ 5|xy$$

which has no solution by Lemma 7.3.2 (v). In the second possibility, we get an equation of the form in Lemma 7.3.2 (*iv*) which has no solution. So we conclude that $k\neq 6,7$ and if $k=8$ we have $P(\Delta_a)=7$.

Let $k=8$. Then 7 divides a_0,a_7. So $P(a_1a_2a_3a_4a_5a_6)\le 5$. By our earlier argument we get $P(a_1a_2a_3a_4a_5a_6)=5$. Since 5 divides exactly two terms, by the argument when (7.4) occurs, we can assume that $5|a_0a_5$ or $5|a_2a_7$. We shall discuss the case $5|a_0a_5$. The other case follows by symmetry. Note that by (7.3) we have $P(a_1a_2a_3a_4)=3$. So we have the following possibilities.

$$(i)\ 3|a_0a_3a_6\ (ii)\ 3|a_1a_4a_7\ (iii) 3|a_2a_5. \tag{7.5}$$

Consider (i) and (ii). Take $[p,q,r,s]=[0,3,4,7]$. This gives rise to

$$a_3a_4(x_3x_4)^{\ell}-a_0a_7(x_0x_7)^{\ell}=12d^2$$

which reduces to

$$Ax^{\ell}+By^{\ell}=d^2 \text{ or } (2d)^2,\ P(AB)\le 5, 7|xy$$

according as $2|\Delta_a$ or not. This equation has no solution by Lemma 7.3.2 (*vii*) for $\ell\ge 11$. Next suppose (*iii*) holds. Note that $\beta_2+\beta_5\equiv 0 \pmod{\ell}$. Consider $[p,q,r,s]=[0,2,5,7]$ which gives rise to

$$a_2a_5(x_2x_5)^{\ell}-a_0a_7(x_0x_7)^{\ell}=10d^2$$

and this reduces to

$$ax^{\ell}+b5^{\gamma}y^{\ell}=d^2,\ \ P(ab)\le 2, \gamma>0, 7|xy \text{ if } 2|\Delta_a$$

and

$$x^\ell + 5^\gamma y^\ell = 2d^2, \gamma > 0, 7|xy \text{ if } 2 \nmid \Delta_a.$$

This has no solution by Lemma 7.3.2 (*vii*) and (*vi*) for $\ell \geq 11$. Thus we are left with $\ell = 7$. We will discuss the three cases in (7.5) one by one.

Case I. $7|a_0a_7;\ 5|a_0a_5;\ 3|a_0a_3a_6$. We have $P(a_1a_2a_4) \leq 2$.

Subcase (a) $7^2|a_0$.
Consider [0, 2, 4]. This is the identity

$$a_0x_0^7 + a_4x_4^7 = 2a_2x_2^7.$$

By Lemma 7.3.1 we get

$$2^{(\alpha_4-\alpha_2-1)} \equiv \pm 1, \pm 18, \pm 19 \pmod{7^2}.$$

Since $|\alpha_4 - \alpha_2 - 1| < 7$, it is easy to check that $\alpha_4 = \alpha_2 + 1$ and by gcd conditions we get

$$(\alpha_2, \alpha_4) \in \{(0, 1), (1, 2)\}.$$

Since $\alpha_4 \geq 1$ we see that

$$2|\Delta_a \text{ and } a_1 = 1.$$

Let $(\alpha_2, \alpha_4) = (0, 1)$. Then $\nu_2(n + 2d) \equiv 0 \pmod 7$ and so

$$a_2 = 1 \text{ and } \nu_2(a_2x_2^7) \geq 7.$$

Consider [1, 2, 3]. This gives rise to

$$x_1^7 + 3^{\beta_3}x_3^7 = 2x_2^7.$$

This has no solution by Lemma 1.6.14. Next let $(\alpha_2, \alpha_4) = (1, 2)$. Then $\alpha_6 = 1$. Consider $[0, 3, 6] \pmod{7^2}$. This gives

$$3^{\beta_3-\beta_6} \equiv \pm 1, \pm 18, \pm 19 \pmod{7^2}$$

giving $\beta_3 = \beta_6 = 1$. Thus

$$a_3 = 3, a_4 = 4, a_6 = 6.$$

Consider [3, 4, 6]. This gives the equation

$$x_3^7 + x_6^7 = 2x_4^7$$

which has no solution by Lemma 1.6.15.

Subcase (b) $7^2|a_7$.
Consider [1, 4, 7] (mod 7^2). We get

$$2^{\alpha_4-\alpha_1+1} \equiv \pm1, \pm18, \pm19 \pmod{7^2}$$

which gives

$$(\alpha_1, \alpha_4) = (1, 0).$$

Thus $a_4 = 1$. Now take [4, 6, 7] (mod 7^2). This gives

$$3^{\beta_6+1} \equiv \pm1, \pm18, \pm19 \pmod{7^2}$$

so that $\beta_6 = 6$. Then $\beta_0 = \beta_3 = 1$ which is a contradiction since $P(b) \leq 2$.

Case II. $7|a_0a_7$; $5|a_0a_5$; $3|a_1a_4a_7$. We have $P(a_2a_3a_6) \leq 2$.

Subcase (a) $7^2|a_0$.
Consider [0, 2, 6] (mod 7^2). This gives

$$3 \times 2^{\alpha_2-\alpha_6} \equiv \pm1, \pm18, \pm19 \pmod{7^2}$$

so that we get

$$\alpha_6 - \alpha_2 \in \{3, -4\}.$$

This shows that 2 divides a_2 and a_6. Hence

$$a_3 = 1.$$

Now take [0, 3, 6] (mod 7^2) which gives

$$2^{\alpha_6-1} \equiv \pm1, \pm18, \pm19 \pmod{7^2}$$

so that $\alpha_6 = 1$ and hence $\alpha_2 = 5$. Finally consider [1, 2, 3] to get

$$3^{\beta_1}x_1^7 + x_3^7 = 2^6x_2^7$$

which has no solution by Lemma 1.6.14.

Subcase (b) $7^2|a_7$.
Consider [2, 6, 7] (mod 7^2). This gives

$$5 \times 2^{\alpha_6-\alpha_2} \equiv \pm1, \pm18, \pm19 \pmod{7^2}.$$

This implies

$$\alpha_2 - \alpha_6 =\in \{3, -4\}.$$

Thus

$$a_3 = 1.$$

Now take $[3, 6, 7] \pmod{7^2}$ which gives

$$2^{\alpha_6+2} \equiv \pm 1, \pm 18, \pm 19 \pmod{7^2}$$

so that $\alpha_6 = 5$. This means $\alpha_2 = 1$. This is not possible since $4||(a_2x_2^7 - a_6x_6^7)$.

Case III $7|a_0a_7$; $5|a_0a_5$; $3|a_2a_5$. We have $P(a_1a_3a_4a_6) \leq 2$.

Subcase (a) $7^2|a_0$.
Consider $[0, 1, 3] \pmod{7^2}$. This gives

$$3 \times 2^{\alpha_1-\alpha_3} \equiv \pm 1, \pm 18, \pm 19 \pmod{7^2}$$

so that we get

$$\alpha_3 - \alpha_1 \in \{3, -4\}.$$

This shows that

$$a_4 = a_6 = 1.$$

Now consider $[0, 4, 6] \pmod{7^2}$ which gives $3^6 \equiv 2^6 \pmod{7^2}$ which is not possible.

Subcase (b) $7^2|a_7$.
Consider $[1, 3, 7] \pmod{7^2}$. This gives

$$3 \times 2^{\alpha_3-\alpha_1-1} \equiv \pm 1, \pm 18, \pm 19 \pmod{7^2}$$

so that

$$\alpha_3 - \alpha_1 - 1 \in \{3, -4\} \text{ and hence } a_4 = a_6 = 1.$$

Now consider $[4, 6, 7] \pmod{7^2}$ which gives $3^6 \equiv 1 \pmod{7^2}$ which is not possible.

This completes the case $\ell = 7$ and thereby the proof of Theorem 7.1.2. □

7.4 Comments

(1) Observe that all along we have considered the equation

$$\Delta(n, d, k) = y^\ell \text{ or } by^\ell$$

with $\gcd(n, d) = 1$. What happens if $\gcd(n, d) > 1$? Does this equation have solution? It seems that this equation may have infinitely many solutions for suitable $\gcd(n, d)$. For example,

$$\Delta(9, 9, 4) = 9 \cdot 18 \cdot 27 \cdot 36 = 54^3.$$

There can be solutions even when $\gcd(n, d) = 1$ as in $\Delta(1, 24, 3) = 35^2$, $\Delta(1, 1, 4) = 3 \cdot 2^3$, $\Delta(1, 1, 6) = 5 \cdot 12^2$. (Why didn't these examples crop up before?) In [10, p. 489], Shorey relaxed the gcd condition to $d \nmid n$ to show that the above equation implies that k is bounded by a number depending only on ℓ and $\omega(d)$.

(2) In [5] (see Theorem 7.1.1 for $k = 6$), the equation

$$\Delta'(n, 1, k) = y^\ell, k \geq 4$$

was completely solved where $\Delta'(n, 1, k)$ denotes the product $\Delta(n, 1, k)$ with one term omitted. A natural question is what we can say about $\Delta'(n, d, k), d > 1$. In [5, Theorem 2] (for $k > 8$) and [11] (for $k \leq 8$), it was shown that

$$\Delta'(n, d, k) = y^\ell, k \geq 4, d > 1, \ell \geq 3$$

implies that $D_1 > 1$ where D_1 is the maximal divisor of d whose prime divisors are $\equiv 1 \pmod{\ell}$. This is analogous to Theorem 6.1.1. Now one would like to solve this equation completely for small values of k. When $k = 3$, Győry [12] used GFE to solve it. The reader can easily see this now. In [13], this was answered for $4 \leq k \leq 8$. The question on integral solutions involving the product $\Delta(n, d, k)$ or $\Delta'(n, d, k)$ can be translated to questions about *rational* solutions involving the product $x(x + 1) \cdots (x + k - 1)$ where x is a rational variable. This approach enables one to apply the methods of Generalized Fermat Equation more efficiently. We will take up this approach in Chap. 8 in detail.

Exercise 7

(1) Prove the analogue of Theorem 7.1.1 for $k = 7, 8$.

(2) Can you prove the following analog of Fermat's result for polynomials? Suppose f, g and h are relatively prime non-trivial polynomials in $\mathbb{C}[X]$, not all constant. Then

$$f(X)^n + g(X)^n \neq h(X)^n$$

for any integer $n > 2$. What about $n = 2$?

References

1. C.L. Siegel, Über einige Anwendungen diophantischer approximation. Abh. Preuss. Akad. Wiss. Phys-Math. Kl. 1929, Nr. 1, 70 pp. Gesammelte Abhandlungen, vol. I (Springer, Berlin, 1966). [7]
2. G. Faltings, Endlichkeitssätze für abelsche Varieäten über Zahlkörpern. Invent. Math. **71**, 349–366 (1983). [**5, 7**]
3. K. Győry, On the Diophantine equation $n(n+1)\cdots(n+k-1) = bx^{\ell}$. Acta Arith. **83**, 87–92 (1998). [**5, 7**]
4. J.W. Sander, Rational points on a class of superelliptic curves. J. Lond. Math. Soc. **59**, 422–434 (1999). [**7, 8**]
5. N. Saradha, T.N. Shorey, Almost perfect powers in arithmetic progression. Acta Arith. **138**, 363–388 (2001). [**6, 7**]
6. M.A. Bennett, C. Skinner, Ternary Diophantine equations via Galois representations and modular forms. Canad. J. Math. **56**, 23–54 (2004). [**7**]
7. M.A. Bennett, N. Bruin, K. Győry, L. Hajdu, Powers from products of consecutive terms in arithmetic progression. Proc. Lond. Math. Soc. **92**(3), 273–306 (2006). [**7, 8**]
8. K. Győry, L. Hajdu, N. Saradha, On the Diophantine equation $n(n+d)\cdots(n+(k-1)d) = by^{\ell}$. Canad. Math. Bull. **47**, 373–388 (2004). [**7, 8**]
9. K. Győry, L. Hajdu, Á. Pintér, Perfect powers from products of consecutive terms in arithmetic progression. Compositio Math. **145**, 845–864 (2009). [**7**]
10. T.N. Shorey, Exponential diophantine equations involving products of consecutive integers and related equations, in *Number Theory*, ed. by R.P. Bambah, V.C. Dumir, R.J. Hans-Gill, Hindustan Book Agency (1999), pp. 463–495; Trends Math. (Birkhäuser, Basel, 2000), pp. 463–495. [**3, 7, 8**]
11. N. Saradha, T.N. Shorey, On the equation $n(n+d)\cdots(n+(i_0-1)d)(n+(i_0+1)d)\cdots(n+(k-1)d) = y^{\ell}$ with $0<i_0<k-1$. Acta Arith. **129**, 1–21 (2007). [**7**]
12. K. Győry, Power values of products of consecutive integers and binomial coefficients, in *Number Theory and Its Applications*, ed. by S. Kanemitsu, K. Győry (Kluwer, 1999), pp. 145–156. [**7**]
13. P. Das, S. Laishram, N. Saradha, D. Sharma, Rational solutions to the variants of Erdős-Selfridge superelliptic curves. Int. J. Number Theory **19**(7), 1707–1741 (2023). [**7, 8**]

Chapter 8
And Miles to Go ···

$\ell \le e^{3^k}$

Let us recall (5.2).

$$\Delta(n, d, k) = by^{\ell} \text{ in integers } n > 0, d > 1, k > 1, y > 1, b \geq 1, \ell \geq 2, \ell - \text{prime}. \tag{8.1}$$

In the foregone chapters, the reader had seen various attempts by many mathematicians to solve Erdős conjecture, viz.,

Equation (8.1) *with* $k \geq 3$ *implies that k is bounded by an absolute constant.*
And it would have been noticed that the conjecture still remains *open* in its full generality. Shorey [1, p. 490] showed that under *abc*-conjecture Erdős conjecture holds for $\ell \geq 4$. This is refined in [2, Theorem 7] by showing that there are only finitely many possible values for all other parameters n, d, b, y, ℓ. We shall prove these results in Theorem 8.1.1. We shall in fact show in Sect. 8.4 that under an equivalent version of *abc*-conjecture due to Granville and Tucker, Erdős conjecture holds for $\ell \geq 2$. A stronger conjecture states the following.

Equation (8.1) *with* $k \geq 3$ *implies that* $(k, \ell) \in \{(3, 2), (3, 3), (4, 2)\}$.
It is known that (8.1) has infinitely many solutions in all the three cases mentioned above. It was conjectured by Tijdeman [3] that

the number of triples (n, d, k) *satisfying* (8.1) *with* $k > 2, \ell > 1, k + \ell > 6$ *is finite.* Darmon and Granville [4, p. 520] conjectured that

$$\text{Equation (8.1) with } b = 1 \text{ implies that } (k, \ell) = (3, 2).$$

In the case $(k, \ell) = (3, 2)$ we get parametric solutions given by

$$(n, d) \in \left\{ \left((t^2 + 2tu - u^2)^2, 4tu(u^2 - t^2)\right), \left(2(t^2 - u^2)^2, 6t^2u^2 - t^4 - u^4\right) \right\}$$

S. Natarajan, *Perfect Powers—An Ode to Erdős*, Infosys Science Foundation Series,
https://doi.org/10.1007/978-981-96-2599-4_8

with $\gcd(t, u) = 1$ and $t + u$ odd. Euler as early as 1780 proved that there are only trivial four term arithmetic progression whose product is a square, so that $(k, \ell) \neq (4, 2)$ if $b = 1$. Further he showed that there are only trivial integer solutions to $x^3 + y^3 = 2z^3$ which implies that there are no three term arithmetic progressions whose product is a cube. Thus $(k, \ell) \neq (3, 3)$ if $b = 1$. Recently, Bennett and Siksek [5] showed that if $k \geq k_0$ where k_0 is an effectively computable absolute constant then any solution to (8.1) is either $y = 0$ or $d = 0$ or

$$\ell \leq e^{10k}.$$

Thus under Erdős conjecture, ℓ is also bounded. Prior to this, Bennett and Siksek [6] considered

$$\Delta(x + 1, 1, k) = (x + 1) \cdots (x + k) = y^\ell \tag{8.2}$$

with $x, y \in \mathbb{Q}$, $k \geq 3$ and $\ell \geq 2$, ℓ prime. For any fixed k and ℓ this equation can be looked upon as a superelliptic curve of genus at least $(\ell - 1)(k - 2)/2$. When $\ell + k > 6$, the genus exceeds 1, and the number of rational points (x, y) is finite by Faltings' theorem. Thus when $(k, \ell) \notin \{(3, 2), (4, 2), (3, 3)\}$, then (8.2) has only finitely many rational solutions (x, y). Sander [7] made the following conjecture with corrections pointed out in [8, p. 277].

Conjecture. The only rational solutions of (8.2) *with* $k \geq 2$ *satisfy either* $y = 0$ *or*

$$(x, y, k, \ell) = \left(\frac{a^2}{b^2 - a^2}, \frac{ab}{b^2 - a^2}, 2, 2\right), a \neq \pm b$$

or

$$(x, y, k, \ell) = \left(\frac{(1 - 2j)}{2}, \frac{\pm 1}{2^j} \prod_{i=1}^{j} (2i - 1), 2j, 2\right)$$

or

$$(x, y, k, \ell) = (-4/3, 2/3, 3, 3) \text{ and } (-2/3, -2/3, 3, 3)$$

where a, b *and* $j > 0$ *are integers suitably chosen.*

Sander [7] proved the conjecture for $2 \leq k \leq 4$, Lakhal and Sander [9] dealt with the case $k = 5$. Now we know the complete answer for $2 \leq k \leq 34$, see [10]. By the conjecture (8.2) does not have any rational solution if $\ell \geq 4$. A famous theorem of Schinzel and Tijdeman [11] asserts that the equation

$$f(x) = y^\ell$$

satisfies either $y \in \{0, \pm 1\}$ or $\ell \leq \ell_0$ for some effectively computable number $\ell_0 = \ell_0(f)$ where $f(X) \in \mathbb{Z}[X]$ has at least two distinct roots. Analogous to Schinzel and Tijdeman theorem, it was shown in [6] that

$$\text{if (8.2) holds, then } \ell \leq e^{3^k}.$$

We shall prove this in Theorem 8.1.2. Extensions of this result were given in [12, 13]. Finding rational points on (8.2) has direct implication on finding integer points for (8.1), see Sect. 8.2. It is desirable to improve the double exponential bound for ℓ or even the exponential bound above. Toward this, we show that either $\ell \leq k$ or the denominators of the rational solutions are very large, see Theorem 8.1.3. In the last Sect. 8.8, we discuss perfect powers which occur at other places such as Recurrence sequences, Nagell-Ljunggren equation, Lebesgue-Ramanujan-Nagell equation and power sums.

8.1 Statement of the Results

Our first theorem is an application of *abc*-conjecture in which we combine results from [1, 2] mentioned in the introduction.

Theorem 8.1.1 *Under abc-conjecture,* (8.1) *with* $\ell \geq 4$ *implies that there are only finitely many possible values for* n, d, k, b, y, ℓ.

Next, we prove the result of Bennett and Siksek, viz.,

Theorem 8.1.2 *If* (8.2) *holds, then* $\ell \leq e^{3^k}$.

We improve the double exponential bound for ℓ with the aid of Lemma 6.4.4. It is a refined version of [14, Theorem 1.1]. Recall that for any integer $n > 1$, we write

$$n = n_1 n_2$$

where n_1 is the maximal divisor of n such that every prime divisor of n_1 is congruent to $1 \pmod{\ell}$. Consequently, every prime divisor of n_2 is incongruent to $1 \pmod{\ell}$ and $\gcd(n_1, n_2) = 1$. We call n_1 as the ℓ-part of n.

Theorem 8.1.3 *Let* $k \geq 3$ *and* $\ell \geq 11$. *Suppose* (8.2) *has a positive rational solution* (x, y) *with* $y > 1$. *Let* x_ℓ *and* y_ℓ *denote the* ℓ*-part of the denominators of* x *and* y. *Then either*

$$\ell \leq k$$

or

$$x_\ell \geq \theta k^{\ell(1-\ell'/\ell-\ell'/\ell^2)} \text{ and } y_\ell \geq \theta k^{k(1-\ell'/\ell-\ell'/\ell^2)}$$

where ℓ' *is given by*

$$\ell' = \begin{cases} 8 \text{ if } 11 \leq \ell \leq 23 \\ 9 \text{ if } \ell = 29 \\ \lceil 7\ell^{1/4} \rceil \text{ if } \ell \geq 31. \end{cases} \tag{8.3}$$

8.2 Connection Between Rational Solutions of (8.2) and Integral Solutions of (5.2)

Let us take the Eq. (8.2). Suppose

$$\ell > k.$$

Let (x, y) be a rational solution. Assume without loss of generality that the denominators of x and y are positive. Write

$$x = \frac{n}{s} \text{ and } y = \frac{m}{t},$$

with $m \neq 0, s, t$ positive integers and $\gcd(n, s) = \gcd(m, t) = 1$. Then (8.2) becomes

$$(n + s) \cdots (n + ks) = \frac{s^k m^\ell}{t^\ell}.$$

Since the left-hand side is an integer and $\gcd(n, s) = \gcd(m, t) = 1$, we get $s^k = t^\ell$. As ℓ is a prime $> k$, there exists an integer $u > 1$ such that $s = u^\ell$ and $t = u^k$. Thus (8.2) becomes

$$(n + u^\ell) \cdots (n + ku^\ell) = m^\ell, \gcd(n, u) = 1, u > 1. \tag{8.4}$$

Put $d = u^\ell$ and $z = n + d$. Then

$$\Delta(z, d, k) = z(z + d) \cdots (z + (k - 1)d) = m^\ell, \gcd(z, d) = 1. \tag{8.5}$$

Any solution to (8.5) with $d = u^\ell$ will give rise to a solution for (8.4) which in turn will yield a rational solution to (8.2). Thus we can focus on (8.5) which is our usual equation (5.2). Then by the results mentioned in the beginning of Chap. 7, we may assume that

$$k \geq 35.$$

8.3 *abc*-Conjecture and Proof of Theorem 8.1.1

In 1983, Masser and Osterlé made the following conjecture which is popularly known as *abc-conjecture* as a possible approach to Fermat's Last Theorem. Given any integer $n > 1$, let us call the radical of n denoted by $R(n)$ to be

$$R(n) = \prod_{p|n} p.$$

abc-**conjecture**. Let $\epsilon > 0$ be given. Suppose a, b, c are positive integers with $\gcd(a, b, c) = 1$ and $a + b = c$, then there exists a positive number $\kappa(\epsilon)$ such that

$$c \le \kappa(\epsilon)\big(R(abc)\big)^{1+\epsilon}.$$

Fermat's Last theorem follows easily from this conjecture. (Prove!) In fact, many conjectures in Number Theory follow under *abc*-conjecture. See

http://www.math.unicaen.fr/nitaj/abc.html

for an excellent index to various web resources concerning the *abc*-conjecture.

Proof of Theorem 8.1.1 Let $\ell \ge 4$. Our starting point is from the remarks made after Theorems 4.1.2 and 5.1.3. Let us assume that $k > k_0$ where k_0 is a large absolute constant. We know then that (8.1) implies that

$$d \gg k^{C \log\log k} \tag{8.6}$$

where C is an effectively computable absolute constant. By the fundamental argument of Erdős (which the reader would know by heart by now) we can find at least three A_i's say, A_f, A_g and A_h with $0 \le f < g < h < k$ such that

$$\max(A_f, A_g, A_h) \le k^2.$$

Let us consider the identity $[f, g, h]$, i.e.,

$$(h-g)(n+fd) + (g-f)(n+hd) = (h-f)(n+gd)$$

which gives

$$(h-g)A_f X_f^\ell + (g-f)A_h X_h^\ell = (h-f)A_g X_g^\ell. \tag{8.7}$$

Note that

$$X_f^\ell \le (h-f)(n+gd) < k(n+gd) = kA_g X_g^\ell \le k^3 X_g^\ell$$

giving

$$X_f < k^{3/\ell} X_g < kX_g.$$

Similarly, $X_h < kX_g$. Let

$$\mu = \gcd\big((h-g)A_f X_f^\ell, (g-f)A_h X_h^\ell\big).$$

Note that indeed,

$$\mu = \gcd\big((h-g)A_f, (g-f)A_h\big)$$

since primes dividing X_f and X_h are $> k$ and $\gcd(X_f, X_h) = 1$. So we see that $\mu \leq k^3$. Dividing both sides of (8.7) by μ, we get an equation of the form

$$a + b = c$$

where $a = \frac{(h-g)A_f}{\mu} X_f^\ell$, $b = \frac{(g-f)A_h}{\mu} X_h^\ell$, $c = \frac{(h-f)A_g}{\mu} X_g^\ell$ and they are pairwise coprime. Note that

$$R\big((h-g)(g-f)(h-f)A_f A_g A_h\big) \leq k^9$$

since $(h-g), (g-f), (h-f)$ are all $< k$ and A_f, A_g, A_h are $\leq k^2$. So applying *abc*-conjecture, we get for any $\epsilon > 0$,

$$\begin{aligned} X_g^\ell &\leq \frac{(h-f)A_g}{\mu} X_g^\ell \\ &\leq \kappa(\epsilon) k^{9+9\epsilon} \Bigg(\prod_{p \mid X_f X_g X_h} p \Bigg)^{1+\epsilon} \\ &\leq \kappa(\epsilon) k^{9+9\epsilon} \big(k^2 X_g^3\big)^{1+\epsilon} \\ &= \kappa(\epsilon) k^{11+11\epsilon} X_g^{3+3\epsilon} \end{aligned} \tag{8.8}$$

giving

$$X_g \leq \kappa(\epsilon)^{1/(\ell-3-3\epsilon)} k^{(11+11\epsilon)/(\ell-3-3\epsilon)}.$$

Let us fix $\epsilon = 1/6$. Then for $\ell \geq 4$,

$$X_g \leq \kappa_0 k^{25.7} \leq k^{27}$$

by taking $k_0 > \kappa_0$. Thus by (8.8),

$$n + gd = A_g X_g^\ell \leq \kappa_0 k^{15} k^{27(3.5)} < k^{110}. \tag{8.9}$$

In particular, since $g \neq 0$, we get

$$d < k^{110}.$$

Together with (8.6), this means

$$\log\log k \ll 1$$

giving $k \leq k_1$ where k_1 is an absolute constant. Thus $k \leq \max\{k_0, k_1\}$. This proves Erdős conjecture. Since k is bounded, by (8.9), it follows that n and d are also bounded and hence $\Delta(n, d, k)$ is bounded. Therefore by (8.1), b, y and ℓ are also bounded. □

8.4 Alternate Version of *abc*-Conjecture

Note that we could prove Theorem 8.1.1 only for $\ell \geq 4$. What about $\ell = 2, 3$? To tackle these cases as well, an argument was given by Granville using an equivalent version of *abc*-conjecture see [15]. We state here this equivalent form of *abc*-conjecture and show his argument.

Equivalent version of *abc-conjecture. Let $F(X, Y)$ be a binary form. Assume that F has pairwise, non-proportional linear factors in its factorization over $\mathbb{C}$. Given $\epsilon > 0$, there exists a computable constant κ_ϵ depending only of F and ϵ such that if m and n are coprime integers, then*

$$\prod_{p \mid F(m,n)} p \geq \kappa_\epsilon (\max(|m|, |n|))^{\deg F - 2 - \epsilon}.$$

The above version was due to Granville and Tucker, see [16].

Proof of Erdős conjecture using the equivalent *abc*-conjecture. Assume that k is large. As usual, we write

$$n + id = A_i X_i^\ell, 0 \leq i < k, T(A) = \{A_0, \ldots, A_{k-1}\}.$$

Then

$$X_i \leq (n + (k-1)d)^{1/\ell} A_i^{-1/\ell}, 0 \leq i < k. \tag{8.10}$$

For every prime $p < k$, $p \nmid d$, let i_p be such that p appears to the maximum power in the term $n + i_p d$. Let

$$I = \{i_p : p < k, p \nmid d\}.$$

By Erdős argument there exists a subset $S \subseteq T(A) \setminus I$ with $|S| \geq k - \pi(k)$, such that

$$\prod_{A_i \in S} A_i < k^k.$$

Let us divide the interval $[0, k-1]$ into sub-intervals $I_j = [20j, 20j + 19]$, $j \geq 0$. Let

$$J = \{j_1, j_2, \cdots\}$$

where

$$\frac{k-1}{100} < j_1 < j_2 < \cdots < \frac{k-1}{20}$$

be those indices such that

$$I_{j_\alpha} \cap I = \emptyset, \alpha \geq 1.$$

Thus

$$|J| \geq \frac{k-1}{25} - \pi(k) \geq \frac{k}{26}$$

for large k. Let

$$\Phi_\alpha = \prod_{i \in I_{j_\alpha}} A_i, \alpha \geq 1.$$

Then

$$\prod_{1 \leq \alpha \leq |J|} \Phi_\alpha \leq \prod_{A_i \in S} A_i < k^k.$$

So there exists an index $j_{\alpha_0} \in J$ such that

$$\Phi_{\alpha_0} < k^{k/|J|} \leq k^{26}. \tag{8.11}$$

Let

$$H = \prod_{i \in I_{j_{\alpha_0}}} (n + id) = \prod_{i=0}^{19} n + (20 j_{\alpha_0} + i)d.$$

Then H is a binary form

$$H = \prod_{i=0}^{19} (X + iY), X = n + 20 j_{\alpha_0} d, Y = d, \gcd(X, Y) = 1.$$

So by the equivalent version of *abc*-conjecture, we get

$$\prod_{p|H} p \geq \kappa_\epsilon (n + 20 j_{\alpha_0} d)^{18-\epsilon}.$$

Since $j_{\alpha_0} \geq (k-1)/100$, taking $\epsilon = 1$ we find

$$\prod_{p|H} p \geq C(n + (k-1)d)^{17} \tag{8.12}$$

where C is an absolute constant. Let us find an upper estimate for the left-hand side quantity in (8.12). We write

$$\prod_{p|H} p = \prod_{\substack{p|H \\ p \leq k}} p \prod_{\substack{p|H \\ p > k}} p.$$

By (8.10),

$$\begin{aligned}\prod_{\substack{p|H\\p>k}} p \leq & \prod_{i=0}^{19} X_{20j_{\alpha_0}+i}\\ \leq & (n+(k-1)d)^{20/\ell}\left(\prod_{i=0}^{19} A_{20j_{\alpha_0}+i}\right)^{-1/\ell}\\ \leq & (n+k-1)d)^{20/\ell}\Phi_{\alpha_0}^{-1/\ell}.\end{aligned}$$

Next,

$$\prod_{\substack{p|H\\p<k}} p \leq \prod_{i=0}^{19} A_{20j_{\alpha_0}+i} = \Phi_{\alpha_0}.$$

Thus by (8.11),

$$\begin{aligned}\prod_{p|H} p &\leq (n+k-1)d)^{20/\ell}\Phi_{\alpha_0}^{1-1/\ell}\\ &\leq (n+k-1)d)^{20/\ell}k^{26(1-1/\ell)}\end{aligned}$$

which together with (8.12) and $\epsilon = 1$ gives

$$k^{26(1-1/\ell)} \geq C(n+(k-1)d)^{17-20/\ell} \geq Ck^{17\ell-20}$$

since $n+(k-1)d > k^\ell$ and this gives the necessary contradiction even for $\ell \geq 2$. □

8.5 Proof of Theorem 8.1.2

Let us assume that there exists a rational point (x, y) satisfying (8.2) with $y \neq 0$. As seen in Sect. 8.2, this means (8.4) and (8.5) are satisfied. First we prove two lemmas. In the first lemma, we find a ternary equation with some nice properties.

Lemma 8.5.1 *Let $k \geq 3$ and ℓ a prime $> k$. Suppose* (8.2) *has rational point (x, y) with $y \neq 0$. Let $3 \leq p \leq k$ be a prime either dividing d or dividing at least one term and at most two terms of $\Delta(z, d, k)$. Then there are non-zero integers a, b, c, x, y, z satisfying*

$$ax^\ell + by^\ell + cz^\ell = 0$$

such that

(1) a, b, c are ℓth power free integers.
(2) $P(abc) < k$
(3) $p \nmid abc$
(4) p divides precisely one of x, y, z.

Proof By the given hypothesis, Eq. (8.4) is valid. Let us first consider the case when $p|d$. Then $p|u$, $p \nmid (n + ju^\ell)$ for $1 \le j \le k$ since $\gcd(n, d) = 1$. We write

$$n + ju^\ell = a_j x_j^\ell \text{ where } a_j \text{ is } \ell\text{-th powerfre and } P(a_j) < k, 1 \le j \le k.$$

Since $p|u$, we see that $p \nmid a_j$, $1 \le j \le k$. Then the identity

$$(n + 2u^\ell) - (n + u^\ell) = u^\ell$$

gives rise to a ternary equation satisfying all the requirements of the lemma.

Now let $p \nmid d$. Suppose p divides only one term of $\Delta(z, k, d)$, say $n + ju^\ell$. Then p occurs to an ℓ-power in $n + ju^\ell$. So $p \nmid a_j$ and $p|x_j$. Take

$$j_0 = \begin{cases} 2 \text{ if } j = 1 \\ k - 1 \text{ if } j = k \\ j + 1 \text{ or } j - 1 \text{ otherwise.} \end{cases}$$

Then

$$(n + ju^\ell) - (n + j_0 u^\ell) = (j - j_0) u^\ell$$

satisfies the requirements of the lemma. Next suppose p divides exactly two terms of $\Delta(z, d, k)$, say, $n + j_1 u^\ell, n + j_2 u^\ell$ with $1 \le j_1 < j_2 \le k$. Then $j_2 = j_1 + p$. Form the identity $[j_1, j_1 + 1, j_1 + p - 1, j_1 + p]$, i.e.,

$$(n + (j_1 + 1)u^\ell)(n + (j_1 + p - 1)u^\ell) - (n + j_1 u^\ell)(n + (j_1 + p)u^\ell) = (p - 1)u^{2\ell}.$$

This gives rise to a ternary equation satisfying the conditions of the lemma.

We appeal to modular forms and Galois representations to derive the following lemma. □

Lemma 8.5.2 *Let $k \ge 3$. Suppose* (8.2) *has rational point (x, y) with $y \ne 0$. Let $3 \le p \le k$ be a prime either dividing d or dividing at least one term and at most two terms in $\Delta(z, d, k)$. Then*

$$\log \ell \le \frac{16\left(\prod_{\substack{q<k, q\ne p \\ q-prime}} q\right) + 1}{6} \log(\sqrt{p} + 1) \le 3^k.$$

Proof As mentioned in Sect. 8.2, we may assume that

$$k \geq 35.$$

Further, we may assume that

$$\ell > k,$$

otherwise, lemma is true. By the given hypothesis, Lemma 8.5.1 is valid. Thus there are non-zero integers a, b, c, x, y, z with

$$ax^\ell + by^\ell + cz^\ell = 0$$

and satisfying (1)–(4). Let $\mu = \gcd(ax^\ell, by^\ell, cz^\ell)$. Dividing the above equation by μ we will get a similar equation with all the properties (1)–(4) in which all the three terms are coprime. So we assume without loss of generality that $\mu = 1$. Further by permuting the three terms and changing signs if necessary, we suppose that

$$ax^\ell \equiv -1 \pmod 4, by^\ell \equiv 0 \pmod 2.$$

Form the Frey curve

$$E : Y^2 = X(X - ax^\ell)(X + by^\ell).$$

Since the prime $p \leq k$ we have $\ell \neq p$. The conductor N_E of this curve is

$$N_E = 2^t R(abc(uvw)^\ell) = 2^t \prod_{\substack{q<k,\\ q\neq p}} q \prod_{q|xyz} q$$

where $t \leq 5$. The minimal discriminant Δ_E is given by

$$\Delta_E = \begin{cases} 2^4(abc)^2(xyz)^{2\ell} \text{ if } v_2(by^\ell) < 4 \\ \frac{(abc)^2(xyz)^{2\ell}}{2^8} \text{ if } v_2(by^\ell) \geq 4. \end{cases}$$

If a prime $q \neq 2$ divides xyz, then

$$v_q(\Delta_E) \equiv 0 \pmod \ell.$$

Hence by Theorem 1.6.15 , $E \sim_\ell f$ for some new form of level N' where

$$N' | 2^4 \prod_{\substack{q<k,\\ q\neq p}} q.$$

By Lemma 8.5.1 (4), E has multiplicative reduction at p and so $p||N_E$. Let K be the associated number field to the newform. Then by Lemma 1.6.11

$$\ell \big| \mathrm{Norm}_{K/\mathbb{Q}}(p + 1 \pm c_p).$$

Further we know that $c_p \neq 0$ and

$$|c_p| \leq 2\sqrt{p}.$$

So

$$\ell \leq (p + 1 + 2\sqrt{p})^{[K:\mathbb{Q}]} = (\sqrt{p} + 1)^{2[K:\mathbb{Q}]}.$$

It is known that

$$[K : \mathbb{Q}] \leq g_0^+(N')$$

where $g_0^+(N')$ denotes the dimension of the space of cuspidal newforms of weight 2 and level N'. By the works of Martin [17], it follows that

$$g_0^+(N') \leq \frac{N' + 1}{12}.$$

Thus by (1.14) and $k \geq 35$, we get

$$\begin{aligned} \log \ell &\leq \frac{16 \prod_{q<k, q \neq p} q + 1}{6} \log(\sqrt{p} + 1) \\ &\leq \frac{16 \times (2.78)^k + 1}{6} \log(\sqrt{k} + 1) \\ &\leq 3^k \end{aligned}$$

which proves the lemma. □

Proof of Theorem 8.1.2. By Bertrand's postulate, there exists a prime p with

$$k/2 < p \leq k.$$

This prime satisfies the hypothesis of Lemma 8.5.2. Hence we get $\ell \leq e^{3^k}$. □

8.6 Proof of Theorem 8.1.3

Let (x, y) be a rational solution of (8.2) and assume that $\ell > k$. Then Eqs. (8.4) and (8.5) are satisfied, so that the denominators of x and y are u^ℓ and u^k. As usual, denote the the ℓ-part of u as u_1. Then the ℓ-part of the denominators of x and y are u_1^ℓ and

u_1^k, respectively. We may also assume that $k \geq 35$. By Lemma 6.4.4 with $D_1 = u_1^\ell$, we have $\beta_0 > 1$ and so

$$x_\ell = u_1^\ell \geq \theta k^{\ell-\ell'-\ell'/\ell} \tag{8.13}$$

where ℓ' is as given in (8.3). Thus

$$u_1 \geq \theta^{1/\ell} k^{1-\ell'/\ell-\ell'/\ell^2}.$$

Since $\theta < 1$, we get

$$y_\ell = u_1^k \geq \theta^{k/\ell} k^{k(1-\ell'/\ell-\ell'/\ell^2)} \geq \theta k^{k(1-\ell'/\ell-\ell'/\ell^2)}.$$

From this and (8.13), the theorem follows. □

Note. Take $k = 35$ and $\ell = 37$. Then by Theorem 8.1.3 we get

$$x_\ell \geq \theta\ 35^{18.5};\ y_\ell \geq \theta\ 35^{17.5}$$

showing that the denominators of a rational solution are very large.

8.7 Comments

(1) One can consider a more general Fermat-type equation as

$$Ax^m + By^n = Cz^r \tag{8.14}$$

where A, B, C are fixed non-zero integers, x, y, z, m, n, r are integers > 1 with $\gcd(x, y, z) = 1$. Darmon and Granville [4] showed that this equation has only finitely many solutions in x, y, z provided

$$\frac{1}{m} + \frac{1}{n} + \frac{1}{r} < 1$$

by using a method of Faltings. In this case, there are five 'small' solutions, viz.,

$$1 + 2^3 = 3^2;\ 2^5 + 7^2 = 3^4;\ 7^3 + 13^2 = 2^9;$$

$$2^7 + 17^3 = 71^2;\ 3^5 + 11^4 = 122^2$$

and five large solutions

$$17^7 + 76721^3 = 21063928^2;\ 1414^3 + 2213459^2 = 65^7;\ 33^8 + 1549034^2 = 15613^3;$$

$$43^8 + 96222^3 = 30042907^2;\ 9262^3 + 15312283^2 = 113^7.$$

The large solutions are due to Beukers and Zagier. Note that in each of these solutions, a square is involved. This led to the following conjecture that

there are no integers $m > 2, n > 2, r > 2$ *and coprime positive integers* x, y, z *such that* $x^m + y^n = z^r$.

When

$$\frac{1}{m} + \frac{1}{n} + \frac{1}{r} > 1,$$

then (m, n, r) is a permutation of one of the triples $(2, 2, r)$, $(2, 3, 3)$, $(2, 3, 4)$, $(2, 3, 5)$ and it is known that there are infinitely many solutions in each case. If

$$\frac{1}{m} + \frac{1}{n} + \frac{1}{r} = 1,$$

then (m, n, r) is a permutation of one of the triples $(2, 3, 6)$, $(2, 4, 4)$, $(3, 3, 3)$ and it is known that there are no solutions. See [4].

(2) A result analogous to Theorem 8.1.2 has been shown in [12] (also see [18]) when $\Delta(n, d, k)$ is replaced by $\Delta'(n, d, k)$ where $\Delta'(n, d, k)$ denotes the product $\Delta(n, d, k)$ with one term omitted. For results when more terms are omitted, see [13, 19].

(3) Sometimes one can transform the equation into an elliptic equation by a suitable rational substitution and information about the elliptic curves can be used to solve the original equation. We illustrate this by an example. See [8] and also [20] for many more examples. Let us consider

$$\Delta(n, d, 4) = y^2,$$

i.e.,

$$n(n + d)(n + 2d)(n + 3d) = y^2.$$

Put $X = 2(n + 3d)/n$ and $Y = 6y/n^2$. Then the equation becomes

$$X(X + 1)(X + 4) = Y^2.$$

This corresponds to curve 24A in Cremona's table which has rank 0. The torsion points for this curve are $\{(0, 0), (-1, 0), (-4, 0), (-2, \pm 2), (2, \pm 6), \mathcal{O}\}$. The first three points correspond to $n + 3d = 0, n + 2d = 0, n + d = 0$ and the last one corresponds to $n = 0$. When $(X, Y) = (2, \pm 6)$, we get $d = 0$. When

$(X, Y) = (-2, \pm 2)$ we get $(n, d) = (-3, 2)$. Thus the only solution with $d > 0$ is

$$(-3)(-1)(1)(3) = 3^2.$$

Information about rank, torsion points etc., of elliptic curves are available in Magma and many other symbolic computational packages.

8.8 Related Problems

So far, we have dealt with the question of finding perfect powers in products of consecutive integers or terms from an arithmetic progression. This leads to a Diophantine equation which is rich with lot of arithmetical properties. A natural question is where else we can find *perfect powers* that may lead to such profound Diophantine equations. There are well known such equations and we shall describe a few of them. All these equations afford rich history and pose complications in their simplicity! They have attracted the attention of several mathematicians and classical to modern methods are used with intricate analysis to tackle these equations. These equations are considered as superelliptic equations and with the advent of Baker's theory of Linear Forms in Logarithms in late sixties, finiteness of the variables involved were shown. In general, the bounds obtained are enormous to have any practical use. More modern techniques and computational ideas entered the fray. We will only give a *cursory* look at them in the sense that we will describe the context, a couple of proven results and some references wherein the interested reader may get more details.

Perfect Powers in Recurrence Sequences

It is only natural to look for perfect powers in sequences. The sequence of natural numbers is abundant with numbers which are perfect powers. Let us look at the well known *Fibonacci* sequence: 0 1, 1, 2, 3, 5,.... Are there perfect powers in this sequence? In other words, if F_n denotes the nth term of this sequence, then does

$$F_n = y^\ell, \ell \geq 2$$

hold? The answer to this simple question required several classical and modular methods that we have encountered in the previous chapters. We see that $F_0 = 0$, $F_1 = F_2 = 1$, $F_6 = 8$, $F_{12} = 144$ are perfect powers. In 2006, Bugeaud, Mignotte and Siksek [21] showed that there are no more perfect powers in the Fibonacci sequence, i.e.,

$$F_n = y^\ell$$

has no solution except when $n = 0, 1, 2, 6, 12$. Following this, Luca and Shorey [22] considered the equation

$$F_n F_{n+d} \cdots F_{n+(k-1)d} = y^\ell \text{ in integers } n > 1, k \geq 2, d \geq 1,$$

i.e., perfect powers in a product with terms from the Fibonacci sequence and showed that this equation has no solution. Note that $F_1 \cdot F_2 = 1$. In fact, the results of [21, 22] deal with more general sequences like Lucas and Lehmer sequences.

Nagell-Ljunggren Equation

Here is a classical equation

$$\frac{x^n - 1}{x - 1} = y^\ell, x > 1, y > 1, \ell > 1, n > 2. \tag{8.15}$$

We ask for powers with all the digits equal to 1 in their x-adic expansions (i.e., to the base x). This is known as the *Nagell-Ljunggren equation* named after initial contributors to the above equation. Observe that

$$\frac{3^5 - 1}{3 - 1} = 11^2; \frac{7^4 - 1}{7 - 1} = 20^2; \frac{18^3 - 1}{18 - 1} = 7^3.$$

It has been conjectured that these are the only solutions of (8.15). This is still *open.* Ljunggren [23] confirmed this conjecture when $\ell = 2$. A milder conjecture is that this has only finitely many solutions and it is known to follow from *abc*-conjecture. There are several partial results on this equation (see [24]). Bugeaud and Mignotte [25] settled an old conjecture that (8.15) does not hold with $x = 10$. In other words, none of the numbers

$$11, 111, 1111, \ldots$$

is a perfect power. Another interesting result is that (8.15) has no solution if x is a square [26, 27]. For instance, none of the numbers

$$91, 820, 7381, 66430, 597871, \ldots$$

is a perfect power. Let us restrict to y being a prime, x a prime power and $\ell \geq 1$. Such equations arise in group theory while constructing algebraic number fields which are arithmetically equivalent but not isomorphic. The only known solution for $\ell \geq 2$ is $(x, y, \ell, n) = (3, 11, 2, 5)$ and it is the only solution when $\ell = 2$. On the other hand, when $\ell = 1$, there are two solutions known:

$$\frac{8^3 - 1}{8 - 1} = 73 \text{ and } \frac{11^9 - 1}{11^3 - 1} = 1772893.$$

For more information on this aspect, we refer to [28]. We also refer to the expository article of Shorey [1] and the paper of Bennett and Levin [29] for various results on this equation.

Lebesgue-Ramanujan-Nagell equations

In 1913, Ramanujan conjectured that *all* the solutions (x, n) of the equation

$$x^2 + 7 = 2^n$$

are given by

$$(x, n) \in \{(1, 3), (3, 4), (5, 5), (11, 7), (181, 15)\}.$$

Nagell solved it in 1948. See also the paper of Skolem et al. [30] for another proof using Skolem's p-adic method. Since then several variations of this equation were considered. For instance, given an integer D take

$$x^2 + D = y^n, \text{ in positive integers } x, y \text{ and } n. \tag{8.16}$$

In 1850, Lebesgue showed that $D = 1$ has no solution. (Try the case $D = -1$). At present, this equation has been solved for $D \leq 100$ due to the efforts of many mathematicians. See [21]. A result of Siegel [31] says that

If $f(x) \in \mathbb{Z}[x]$ has at least two distinct roots then $P(f(x)) \to \infty$ as $|x| \to \infty$.
It follows that the number of solutions (x, n) of (8.16) is finite provided y is fixed. In 1960, Apéry [32] showed that the equation

$$x^2 + D = 2^{n+2}, D \neq 7$$

has at most two solutions. In particular,

$$(x, n) \in \{(3, 3), (45, 9)\} \text{ if } D = 23$$

and

$$(x, n) \in \{(1, m - 1), (2^m - 1, 2(m - 1))\} \text{ if } D = 2^{m+1} - 1 \text{ with } m \geq 4.$$

Thus there are infinitely many D for which there are exactly two solutions. There is a huge literature on this equation and its generalizations. We refer to [1].

Perfect Powers in Power Sums

It is well known that

$$1^3 + 2^3 + \cdots + k^3 = \left(\frac{k(k+1)}{2}\right)^2,$$

i.e., the sum of the cubes of the first k consecutive positive integers is a perfect square. We can generalize this and look at the equation

$$1^m + 2^m + \cdots + k^m = y^\ell \text{ in integers } k \geq 2, m \geq 1, \ell \geq 1. \tag{8.17}$$

In 1956 Schäffer conjectured that if

$$(\ell, m) \notin \{(1, m), (2, 1), (2, 3), (2, 5)\}$$

then the only solution of (8.17) is

$$(\ell, m, k, y) = (2, 2, 24, 70).$$

The particular case $(y, \ell) = (k+1, m)$ was conjectured by Erdős and Moser. For the four values of (ℓ, m) given above there are infinitely many solutions as shown below which were proved by Schäffer himself.

$$(i)\quad (\ell, m) = (1, m) : 1^m + \cdots + k^m = y \text{ for any } k;$$

$$\begin{aligned}(ii)\quad (\ell, m) = (2, 1) : 1 + \cdots + k = y^2,&\\ k = \frac{x_{2r}-1}{2}, y = \frac{y_{2r}}{2} \text{ where}&\\ x_r + y_r\sqrt{2} = (1+\sqrt{2})^r, r = 1, 2, 3, \ldots&\\ i.e., (k, y) = (1, 1), (8, 6), \cdots&\end{aligned}$$

$$\begin{aligned}(iii)\quad (\ell, m) = (2, 3) : 1^3 + \cdots + k^3 = y^2,&\\ y = \frac{k(k+1)}{2} \text{ for any } k.&\end{aligned}$$

$$\begin{aligned}(iv)\quad (\ell, m) = (2, 5) : 1^5 + \cdots + k^5 = y^2,&\\ k = \frac{p-1}{2}, y = \frac{q(3q^2+1)}{4} \text{ where}&\\ p + q\sqrt{6} = (3+\sqrt{6})(5+2\sqrt{6})^s, s = 0, 1, 2, 3, \ldots&\\ i.e., (k, y) = (1, 1), (13, 1001), \ldots&\end{aligned}$$

For $m \geq 2$ and $m \neq 3, 5$ Voorhoeve, Győry and Tijdeman [33] proved the finiteness of the solutions which can be effectively bounded. The equation attracted lot of interest of late due to the innovations in computational techniques. We urge the reader to look into a thorough survey on this equation in [34].

Exercise 1

(1) Discuss the finiteness of the solutions of various equations found in this chapter under *abc*-conjecture.
(2) Show that the assertion of Theorem 8.1.2 holds even if the right-hand side of (8.2) is replaced by by^{ℓ} with $P(b) \leq k/2$. How can one use Sylvester's theorem?

References

1. T.N. Shorey, Exponential diophantine equations involving products of consecutive integers and related equations, in *Number Theory*, ed. by R.P. Bambah, V.C. Dumir, R.J. Hans-Gill, Hindustan Book Agency (1999), pp. 463–495; Trends Math. (Birkhäuser, Basel 2000), pp. 463–495. [**3, 8**]
2. K. Győry, L. Hajdu, N. Saradha, On the Diophantine equation $n(n+d)\cdots(n+(k-1)d) = by^{\ell}$. Canad. Math. Bull. **47**, 373–388 (2004). [**7, 8**]
3. R. Tijdeman, Diophantine equations and Diophantine approximations, in *Number Theory and Applications (Banff, 1988)* (Kluwer, Dordrecht, 1989), pp. 215–243. [**8**]
4. H. Darmon, A. Granville, On the equations $z^m = F(x, y)$ and $Ax^p + By^q = Cz^r$. Bull. Lond. Math. Soc. **27**, 513–543 (1995). [**8**]
5. M.A. Bennett, S. Siksek, A conjecture of Erdős, supersingular primes and short character sums. Ann. of Math. **191**(2), 355–39 2 (2020). [**8**]
6. M.A. Bennett, S. Siksek, Rational points on Erdős-Selfridge superelliptic curves. Compos. Math. **152**, 2249–2254 (2016). [**8**]
7. J. W. Sander, Rational points on a class of superelliptic curves. J. Lond. Math. Soc. **59**, 422–434 (1999). [**7, 8**]
8. M.A. Bennett, N. Bruin, K. Győry, L. Hajdu, Powers from products of consecutive terms in arithmetic progression. Proc. London Math. Soc. **92**(3), 273–306 (2006). [**7, 8**]
9. M. Lakhal, J.W. Sander, Rational points on the superelliptic Erdős-selfridge curve of fifth degree. Mathematika **50**, 113–124 (2003). [**8**]
10. K. Győry, L. Hajdu, Á. Pintér, Perfect powers from products of consecutive terms in arithmetic progression. Compositio Math. **145**, 845–864 (2009). [**7**]
11. A. Schinzel, R. Tijdeman, On the equation $y^m = P(x)$. Acta Arith. **XXXI**, 199–204 (1976). [**8**]
12. P. Das, S. Laishram, N. Saradha, Variants of Erdős-Selfridge superelliptic curves and their rational points. Mathematika **64**, 380–386 (2018). [**8**]
13. N. Saradha, More variants of Erdős-Selfridge superelliptic curves and their rational points. Indian J. Pure Appl. Math. **50**, 333–342 (2019). [**8**]
14. N. Saradha, Rational points with large denominator on Erdős-Selfridge superelliptic curves. Publ. Math. Debrecen **99**, 317-329 (2021). [**2, 8**]
15. S. Laishram, T.N. Shorey, Perfect powers in arithmetic progression. J. Comb. Number Theory **7**(2), 95–110 (2015). [**8**]
16. A. Granville, T. Tucker, It's as easy as ABC. Not. Am. Math. Soc. **49**, 1224–1231 (2002). [**8**]
17. G. Martin, Dimensions of the spaces of cuspforms and newforms on $\Gamma_0(N)$ and $\Gamma_1(N)$. J. Number Theory **112**, 298–331 (2005). [**8**]
18. S. Edis, On a variation of the Erdős-Selfridge Superelliptic curve. Bull. Lond. Math. Soc. **51**, 633–638 (2019). [**8**]
19. M.A. Bennett, Powers from products of terms in progressions with gaps. Acta Arith. **210**, 205–222 (2023). [**8**]
20. P. Das, S. Laishram, N. Saradha, D. Sharma, Rational solutions to the variants of Erdős-Selfridge superelliptic curves. Int. J. Number Theory **19**(7), 1707–1741 (2023). [**7, 8**]

21. Y. Bugeaud, M. Mignotte, S. Siksek, Classical and modular approaches to exponential and Diophantine equations II. The Lebsgue-Nagell equation. Compos. Math. **142**(1), 31–62 (2006). [**8**]
22. F. Luca, T.N. Shorey, Diophantine equations with products of consecutive terms in Lucas sequences. J. Number Theory **114**, 298–311 (2005). [**8**]
23. W. Ljunggren, Noen Setninger om ubestemte likninger av formen $(x^n - 1)/(x - 1) = y^q$. Norsk. Mat. Tidsskr. 1. Hefte **25**, 17–20 (1943). [**8**]
24. T.N. Shorey, R. Tijdeman, New applications of Diophantine Approximations to Diophantine equations. Math. Scand. **39**, 5–18 (1976). [**8**]
25. Y. Bugeaud, M. Mignotte, On integers with identical digits. Mathematika **46**, 411–417 (1999). [**8**]
26. N. Saradha, T.N. Shorey, The equation $\frac{x^n-1}{x-1} = y^q$ with x square. Math. Proc. Camb. Philos. Soc. **125**, 1–19 (1999). [**8**]
27. Y. Bugeaud, M. Mignotte , M. Roy, T.N. Shorey, The equation $\frac{x^n-1}{x-1} = y^q$ has no solution with x square. Math. Proc. Camb. Philos. Soc., **127**, 353–372 (1999). [**8**]
28. D. Estes, R. Guralnick, M. Schacher, E. Strauss, Equations in prime powers. Pacific J. of Math. **118**, 359–367 (1985). [**8**]
29. M.A. Bennett, A. Levin, The Nagell-Ljunggren equation via Runge's method. Monatsh Math **177**, 15–31 (2015). https://doi.org/10.1007/s00605-015-0748-1. [**8**]
30. T. Skolem, S. Chowla, D.J. Lewis, The Diophantine equation $2^{n+2} - 7 = x^2$ and related problems. Proc. Amer. Math. Soc. **10**, 250–257 (1959). [**8**]
31. C.L. Siegel, Approximation algebraischer Zahlen. Math. Zeit **10**, 173–213 (1921). [**8**]
32. R. Apéry, *Sur une équation diophantienne,* C.R. Acad. Sci. Paris Sér. A **251**, 1263–1264 and 1451–1452 (1960). [**8**]
33. M. Voorhoeve, K. Győry, R. Tijdeman, On the Diophantine equation $1^k + 2^k + \cdots + x^k + R(x) = y^z$. Acta Math. **143**, 1–8 (1979). [**8**]
34. N. Coppola, M. Curćo-Iranzo, M. Khawaja, V. Patel, Ö. Ülkem, Power values of power sums: a survey, arXiv:2306.05168v2 [math.NT], 27 July 2023. [**8**]

Index

S. Natarajan, *Perfect Powers-An Ode to Erdős*, Infosys Science Foundation Series,
https://doi.org/10.1007/978-981-96-2599-4

The manufacturer's authorised representative in the EU is Springer Nature Customer Service Centre GmbH, Europaplatz 3, 69115 Heidelberg, Germany. If you have any concerns regarding our products, please contact ProductSafety@springernature.com

Printed and bound by CPI Group (UK) Ltd, Croydon, CR0 4YY
01/10/2026
02225925-0003